# SPERMIOGENESIS

*DAVID M. PHILLIPS*

*The Population Council*
*The Rockefeller University*
*New York, New York*



*ACADEMIC PRESS New York and London*
*A Subsidiary of Harcourt Brace Jovanovich, Publishers*

ACADEMIC PRESS, INC.
111 Fifth Avenue, New York, New York 10003

*United Kingdom Edition published by*
ACADEMIC PRESS, INC. (LONDON) LTD.
24/28 Oval Road, London NW1

**Library of Congress Cataloging in Publication Data**

Phillips, David M
Spermiogenesis.

(Ultrastructure of cells and organisms)
Bibliography: p.
1. Spermiogenesis in animals. I. Title.
II. Series. [DNLM: 1. Spermatogenesis. WJ384 P558s 1973]
QL966.P43 591.1'6'6 73-18971
ISBN 0–12–553650–X

PRINTED IN THE UNITED STATES OF AMERICA

## CONTENTS

## PREFACE

This monograph is a summary of the process of sperm structure and development in a variety of organisms. Sperm were among the first cells viewed with the microscope. It is easy to imagine why Leeuwenhoek may have selected his own semen as an object for study through a microscope—curiosity about sex is a motive underlying much of human endeavor. Once sperm were discovered, they continued to intrigue microscopists, and studies of odd-looking sperm from organisms of every description were given much attention by early cytologists. Sperm also proved to be gratifying research objects for electron microscopists as they possess beautiful and intricate form at the fine structure level. Electron micrographs of sperm are undeniably awe-inspiring examples of the order and complexity of living things.

Aside from an anesthetic excitement about the beauty of life, a practical motive nowadays enters into the study of sperm. One hopes that a deeper understanding of the process leading to sperm development and sperm function during fertilization will promote new and imaginative approaches to the control of human reproduction.

I hope that this book will be a useful, stimulating reference for students of reproductive biology, and that occasionally the pictures will be viewed with nonscientific naivety to simply admire the incredible and beautiful components of life.

David M. Phillips

## INTRODUCTION

Spermatozoa of almost all species have a few very precise and limited functions—to carry the genetic material to the egg and to introduce the genetic material into the egg. In keeping with the limited and, as far as they are concerned, terminal role that they play in fertilization, sperm are highly specialized cells devoid of most organelles generally associated with ongoing cellular metabolism. The essential elements they retain are packed into precise shapes that must be consistent with motility. Thus, cellular organelles that are retained in sperm are highly modified. The anteriormost region of the spermatozoon is generally coated by a precisely constructed acrosome (Fig. 2). The acrosome is originally derived during spermiogenesis from a single large secretion granule. The nucleus of the spermatozoon underlies the acrosome. The chromatin within the precisely shaped sperm nucleus is highly compacted. Thus, the inactive genetic material is accommodated in a small volume that is presumably favorably shaped for motility. Sperm mitochondria, which transform energy either by catabolyzing substances within themselves or within their surrounding environment, are generally located just behind the nucleus. The number and displacement of the mitochondria within a sperm cell are generally constant for a given species. Spermatozoa are propelled by a flagellum which often contains accessory structures besides the 9 + 2 array of tubules. Sperm generally do not contain organelles other than the acrosome, nucleus, mitochrondria, and flagellum. Unlike most other cells, spermatozoa do not synthesize RNA or protein and lack all of the structures such as ribosomes and nucleoli which are involved in these processes.

The formation of the spermatozoon from the product of the second meiotic division is called spermiogenesis. This process involves an elaborate series of events which have been studied in many species. Although the morphological changes that sperm undergo during spermiogenesis are now well described in a number of species, very little is known about the mechanisms that cause the complex movements and structural alterations of organelles during spermiogenesis.

In this monograph, I shall attempt to illustrate the structure of the organelles that characterize mature spermatozoa and to trace the morphogenesis of these organelles during sperm formation. This task is somewhat complicated by the existence of considerable variation in the structure of spermatozoa among different species and the lack of good morphological studies of sperm structure and spermiogenesis in many animal groups. Most of the illustrations are of insect and mammalian spermatids and sperm since these two classes have been studied more extensively than others. The general morphological features of mammalian sperm structure (see Fig. 1) differ in some ways from the features of insect sperm (see Fig. 29), the spermatozoa produced by insects appearing relatively more simple. When the functions and developmental regions of the organelles are considered, homologies between complex structures found in mammalian sperm and simpler insect sperm structures may be deduced even though the organelles in question are quite dissimilar in appearance. Such homologies will be stressed in an effort to present as general a picture of spermiogenesis as seems consistent with the species-specific variations that exist.

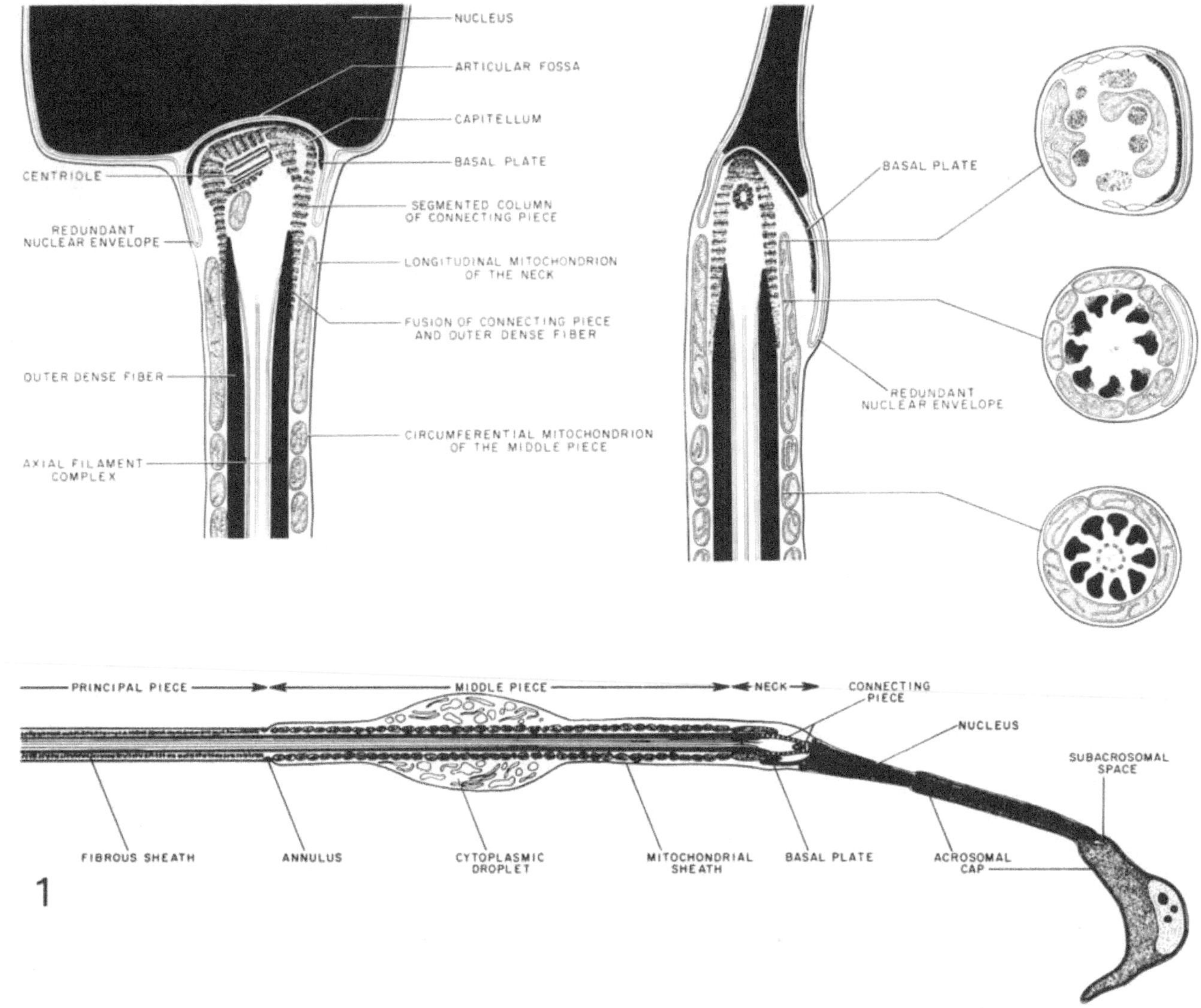


**Fig. 1.** The terminology illustrated here for the parts of the guinea pig spermatozoa has been developed by Fawcett and is applicable to spermatozoa of all mammalian species. Though general similarities exist between spermatozoa of mammals and other animal groups, there are so many differences that most of this terminology is not applied to spermatozoa of other species. From *Z. Zellforsch. Mikrosk. Anat.* **67**, 28 (1965). Courtesy of Dr. D. W. Fawcett.

**Fig. 2.** Longitudinal section of a spermatozoon of the kangaroo rat. Curving of the anterior portion of the midpiece is a characteristic of sperm of this species. Kangaroo rat spermatozoa have a long thin portion of the acrosome extending several microns anteriorly from the electron-dense nucleus. Precisely shaped mitochondria are characteristic of sperm of mammalian species.

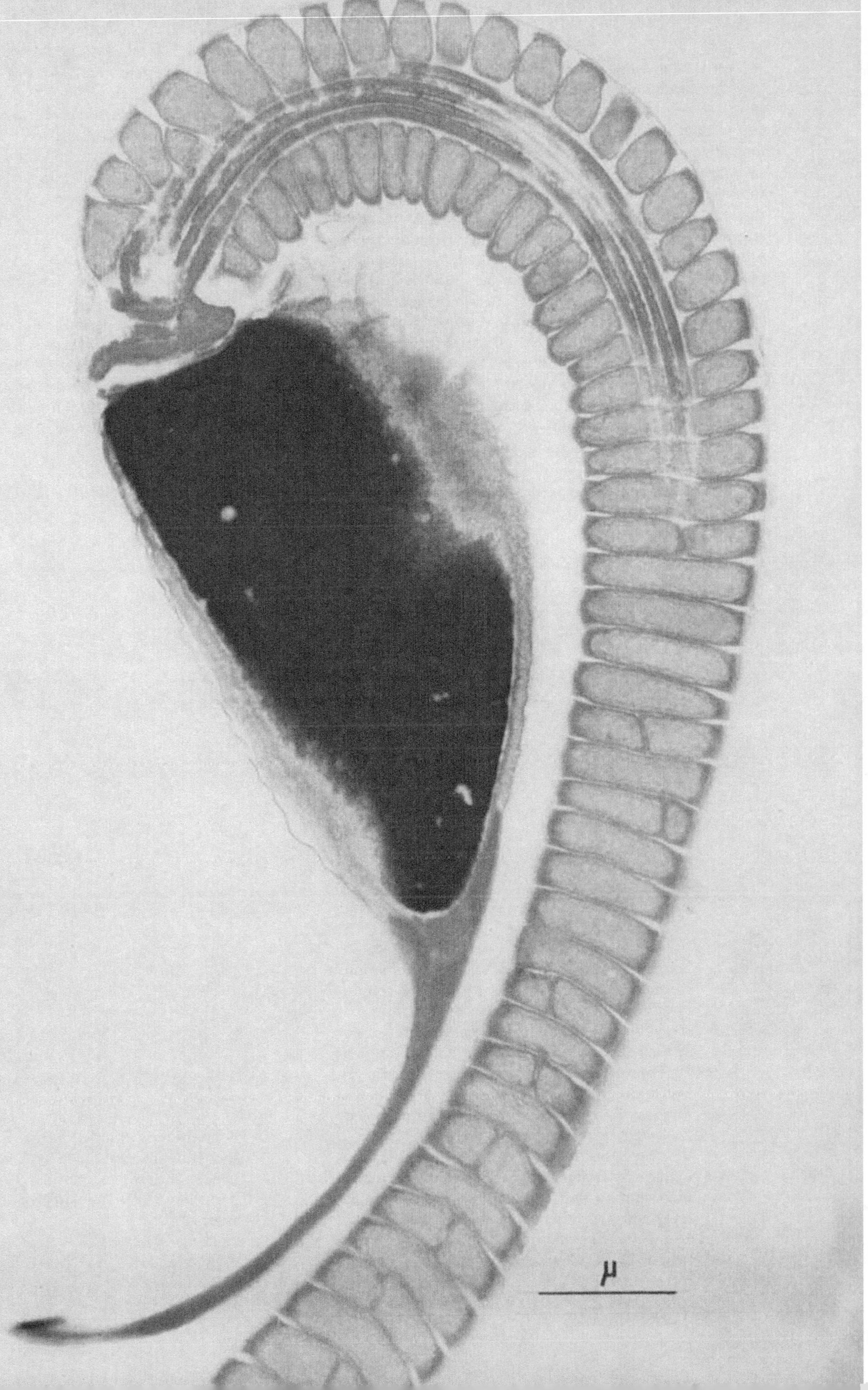


2

## THE SEMINIFEROUS TUBULE

Testes of mammals and most other vertebrates contain hundreds of coiled seminiferous tubules. Each tubule is roughly 100 to 400 μ wide and 5 to 100 cm in length depending on species. Testes are compartmentalized by septi into as many as several hundred compartments, each containing one to five seminiferous tubules. The tubules are composed of a stratified epithelium of germ cells surrounding a lumen. As in other stratified epithelia, cells become increasingly more differentiated as they move away from the basement membrane; that is, spermatogonia are located farthest from the lumen, spermatocytes are more medially located, and spermatids are found nearest the lumen. As they mature, spermatozoa are released into the lumen of the seminiferous tubules. The cells at different stages of spermiogenesis are not randomly distributed in the seminiferous tubule but occur in certain combinations in any given segment of the tubule (*35*). The degree of order varies among species. In the guinea pig testis, Clermont (*35*) has classified 12 types of regions of cellular association, each differing in the combinations of stages of spermiogenesis found there (Fig. 3). In histological cross sections of seminiferous tubules, one can find each of the 12 combinations. Each cell goes through all stages as it moves toward the lumen of the tubule. This process takes 12 days in rat and 48 days in man. If one could look along the length of a seminiferous tubule, all of the stages would be found in order. This process, therefore, is often termed a spermatogenic wave.

A unique type of cell, the Sertoli cell, maintains the integrity of the epithelium within the tubules. Long processes extending from the basally located Sertoli cells surround the germ cells, as can be seen in Fig. 4. Junctions connect the germ cells with the Sertoli cells in places. Sertoli cells were classically thought to serve as a supportive matrix for the germ cells and perhaps to serve a nutritive function as well. Current evidence suggests that the Sertoli cell may be involved in controlling the release of mature sperm into the lumen (*50*).

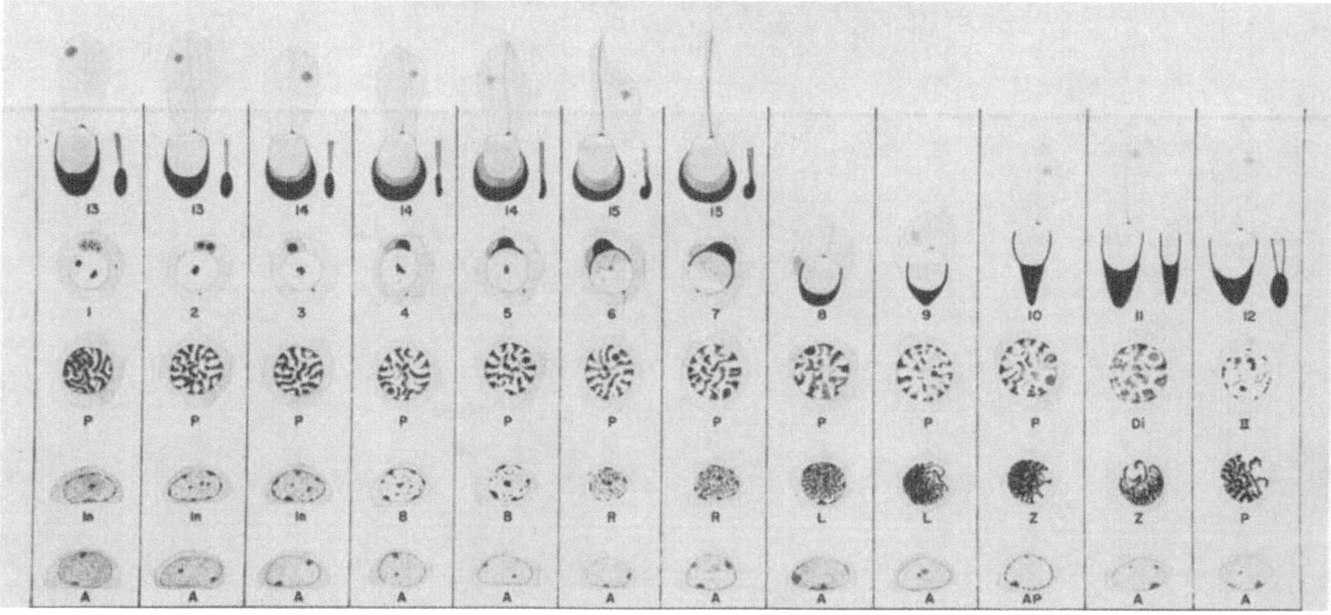

**3**

STAGES OF THE CYCLE

**Fig. 3.** Schematic representation of 12 combinations of male germ cells in the guinea pig seminiferous tubule. A, represents Type A spermatogonia; AP, Type A spermatogonia in prophase; In, intermediate type spermatogonia; B, Type B spermatogonia; R, resting primary spermatocytes; L, leptotene primary spermatocytes; Z, zygotene primary spermatocytes; P, pachytene primary spermatocytes; Di, diakinesis of primary spermatocytes; II, secondary spermatocytes. The changes taking place during spermatogenesis are in sequence from left to right, starting with the bottom row. From *Fert. Steril.* **11**, 563 (1960). Courtesy of Dr. Y. Clermont.

**Fig. 4.** Seminiferous tubule of the Chinese hamster showing processes of Sertoli cell surrounding heads of late spermatids. Cells around the periphery of the micrograph are early spermatids.

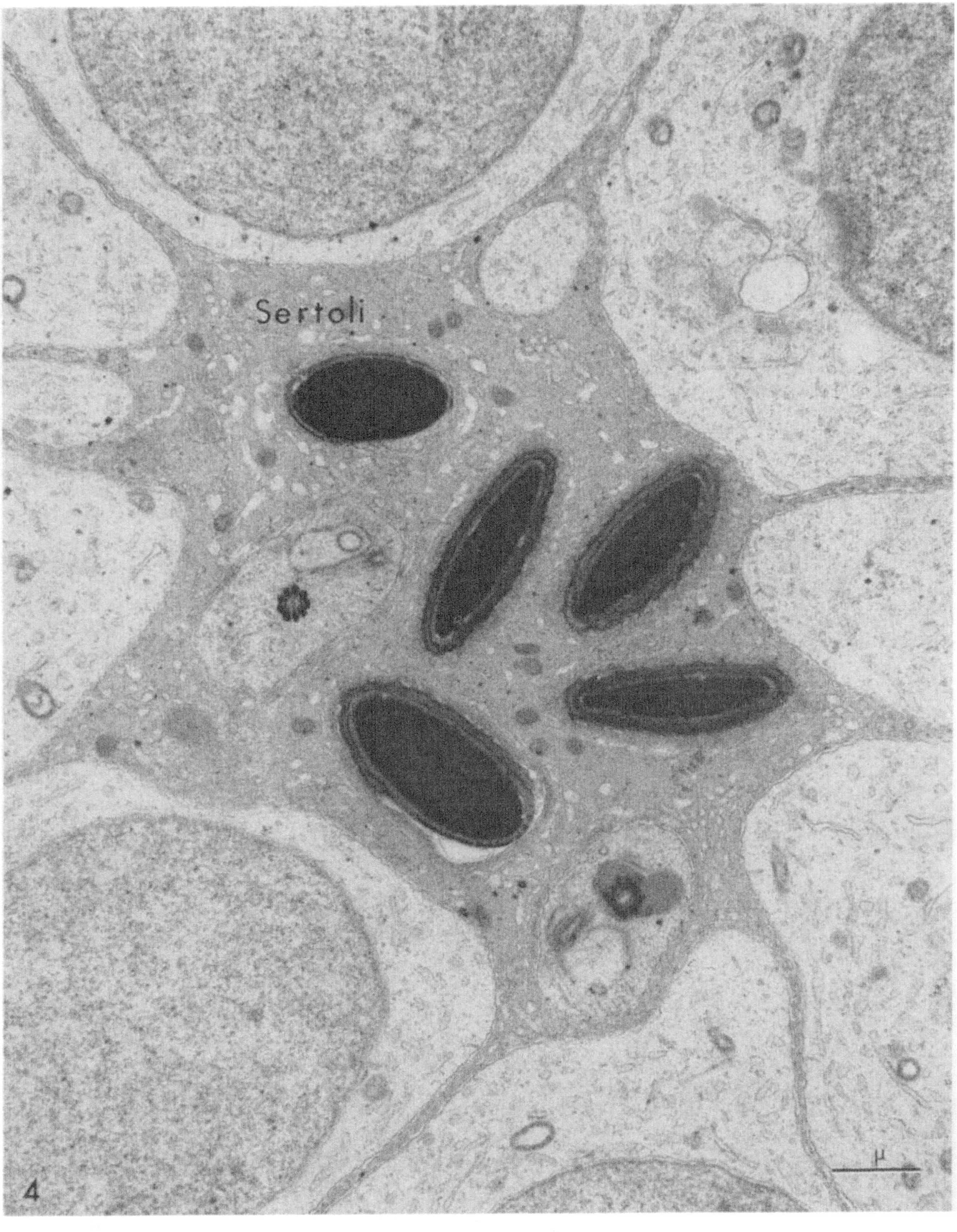
Sertoli
μ
4

## THE CYST

In contrast to the case in most vertebrates in which sperm develop in seminiferous tubules, in most invertebrates and some lower vertebrates, the male germ cells develop in cysts. A cyst is essentially a group of germ cells surrounded by an epithelium. The number of spermatids per cyst is usually constant for a given species and generally equals $2^x$ where $x$ is the number of gonial divisions plus the two meiotic divisions. In Fig. 5, a transverse section through three cysts containing late spermatids of the caddisfly *Platycentropis,* where each black dot is the nucleus of a spermatid, one can see that there are approximately 128 ($2^7$) cells per cyst in this species. Species have been described with characteristically 16, 32, 64, 128, 256, 512, and 1024 spermatids per cyst. Closely related species tend to have like numbers of germ cells per cyst (*145*). As a result of incomplete cytokinesis, the products of late gonial and meiotic divisions in most male animals remain connected by a cytoplasmic bridge that is remnant of the midbody. Thus, the cells of a cyst are functionally a syncytium; they divide and develop in synchrony. In insects, the interconnected spermatids within a cyst remain aligned side by side in register throughout spermiogenesis so that a transverse section through the cyst cuts all cells at the same level. In some insect species (*106, 128, 131*), the whole cyst is released from the testis intact and swims as a unit held together by a sheath of epithelial cells or extracellular material. In most species, however, the cyst breaks open to release the sperm when it is mature.

**Fig. 5.** Three transversely cut cysts of the caddisfly *Platycentropis* showing hexagonally packed late spermatid surrounded by large polytene cells.

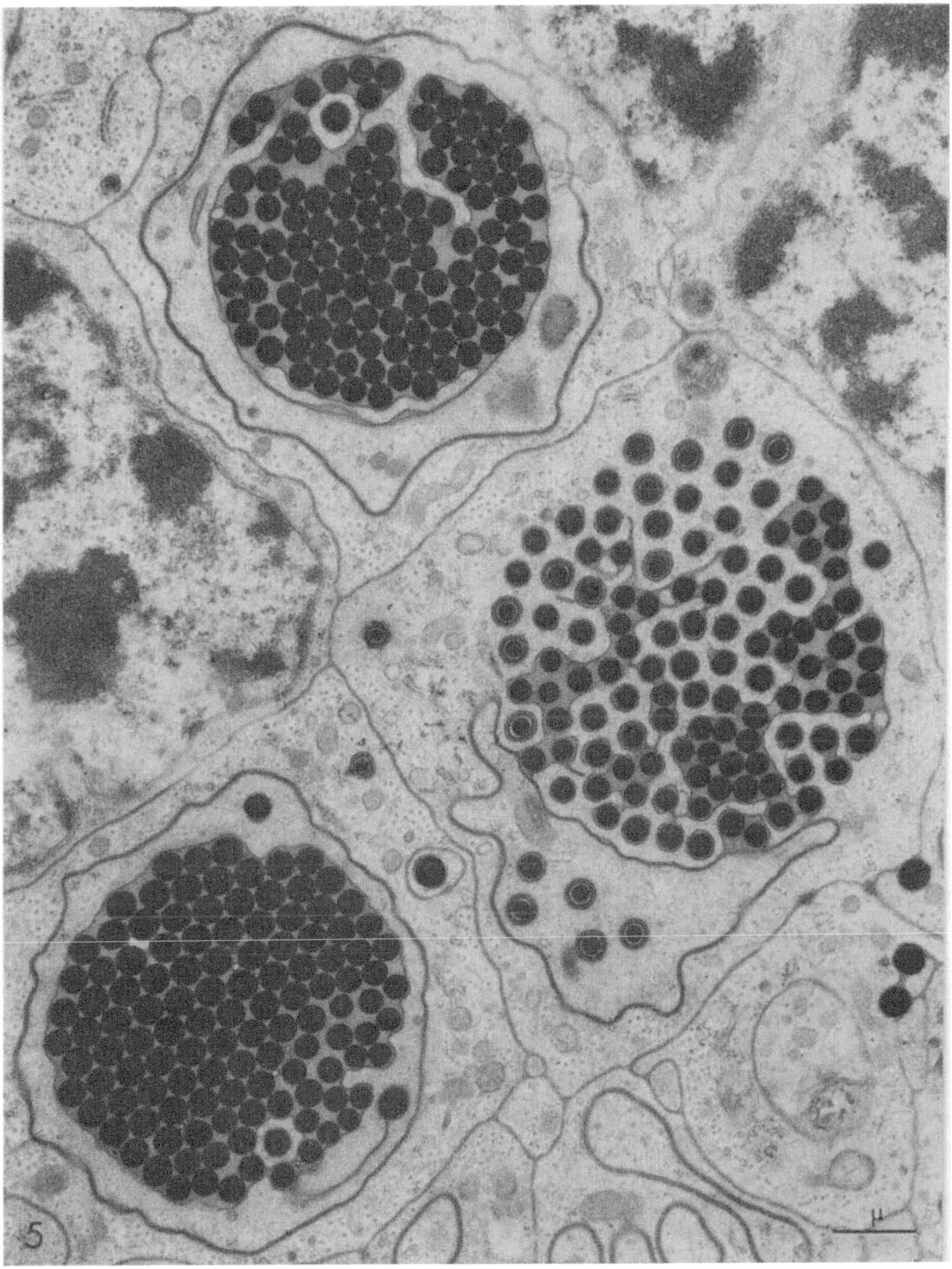
5
μ

## INTERCELLULAR BRIDGES

Discrete groups of set numbers of spermatogonia, spermatocytes, or spermatids connected by intercellular bridges occur in many species throughout the phylogeny. This implies that early in gonial divisions, the cells must complete cytokinesis, but, at some point in development, the product of a gonial division does not separate completely from its sister cell and at that time the remaining number of mitotic divisions that will occur before meiosis is determined (Fig. 6). This situation is strikingly apparent in insects in which spermatids of a cyst are aligned side by side in register (*104*); in some insects, notably among the Diptera, the intercellular bridges are extensive and all at the same level. In favorable sections, it is possible to see bridges between nearly all germ cells of a cyst (*15*). Although the maximum number of spermatids in an interconnected group is not known in mammals, the frequency of such bridges in random sections suggests that large numbers of spermatids, possibly as many as a hundred, may be interconnected (*40*).

In many species, material observed during division remains around the intercellular bridges on the inside of the plasma membrane forming a dense collar. Such material can be seen in the three intercellular bridges connecting this rabbit spermatid to three other spermatids (Fig. 7).

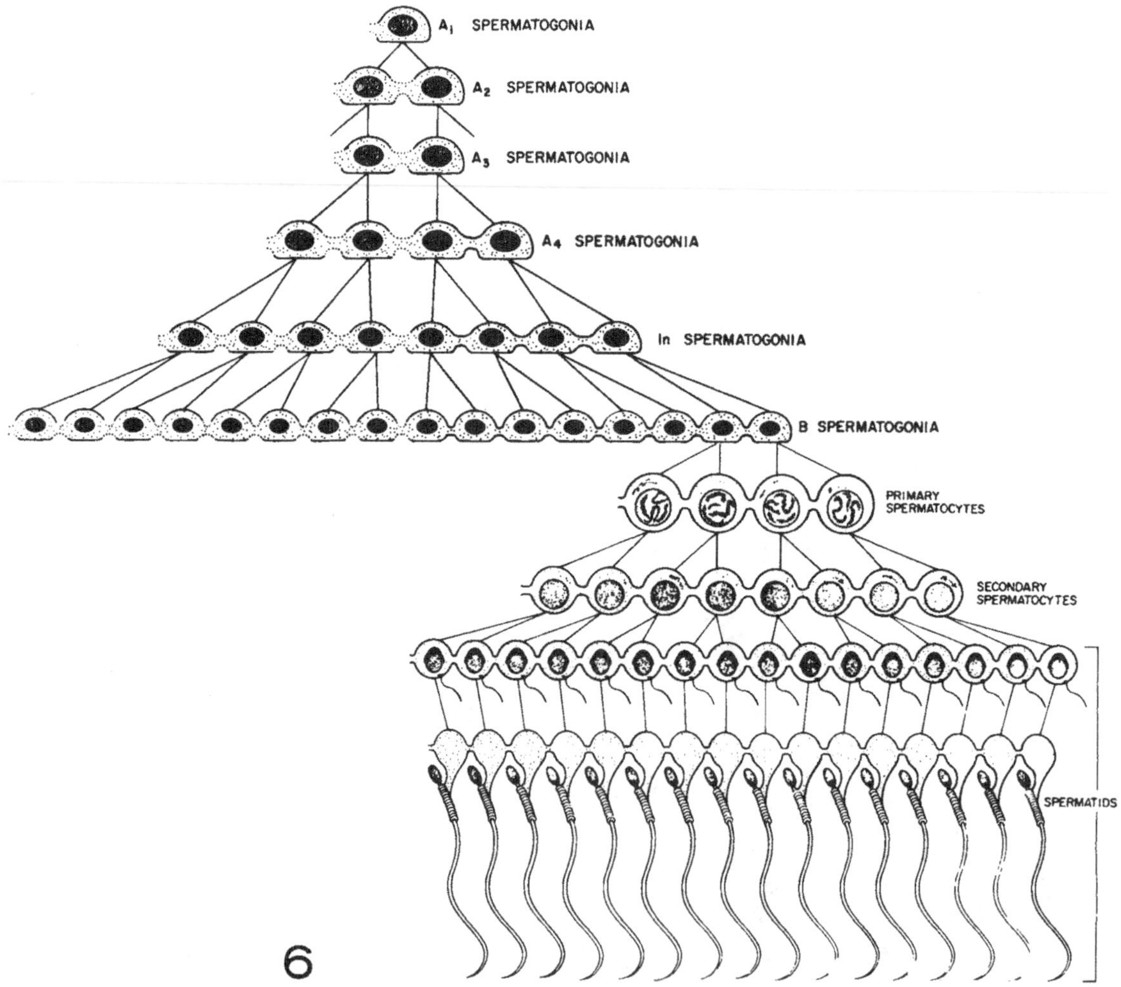


**Fig. 6.** The developmental history of intercellular bridges connecting mammalian spermatogonia, spermatocytes, or spermatids. The large numbers of germ cells with confluent cytoplasm within a group divide and differentiate synchronously (or nearly synchonously). The left side of each row of cells is open indicating that the maximum number of interconnecting mammalian spermatogonia, spermatocytes, and spermatids is not known. From *Biol. Reprod.* 4, 2 (1971). Courtesy of Drs. Dym and Fawcett.

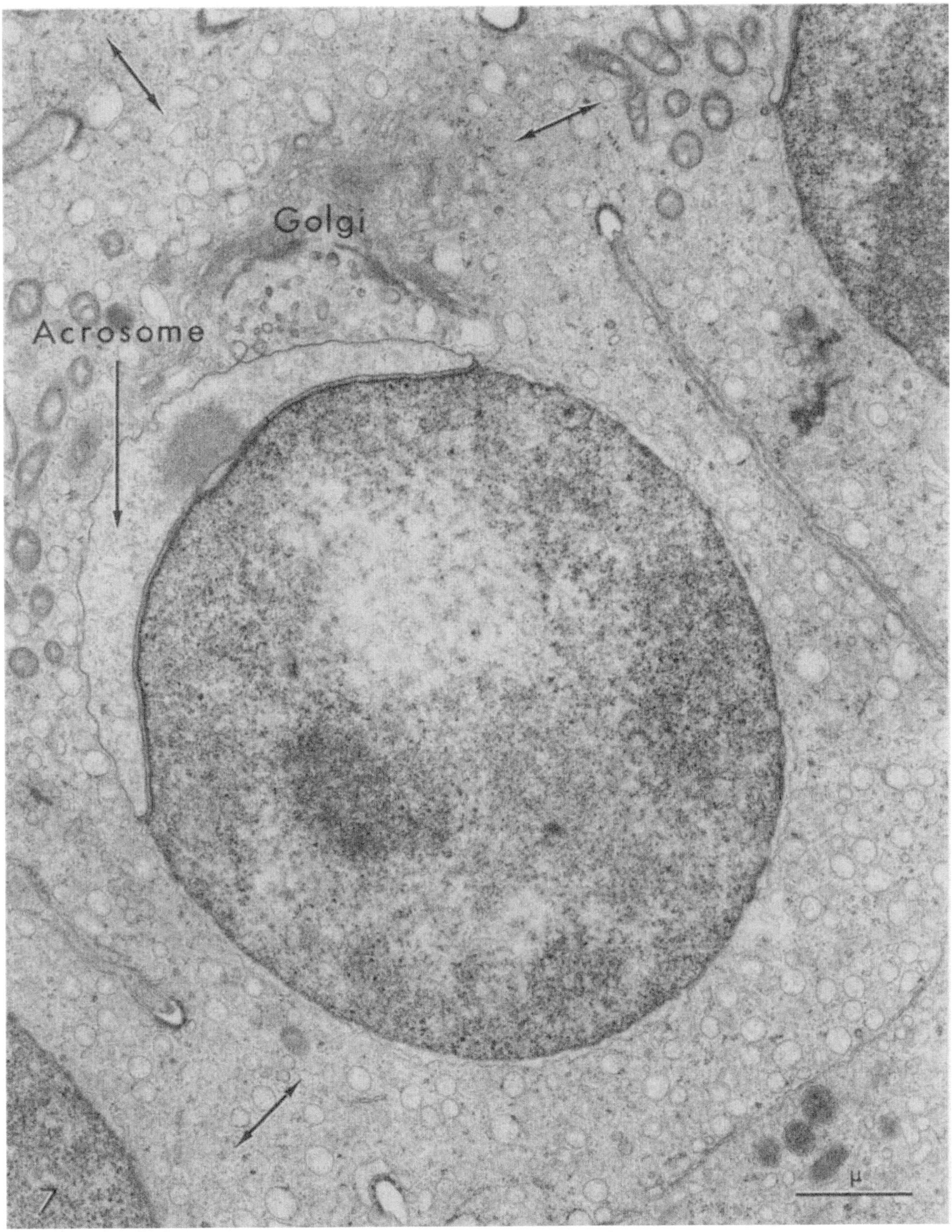


**Fig. 7.** Early rabbit spermatid showing wide bridges connecting three spermatids. The dense material in the center of the developing acrosome is adjacent to the nucleus, a typical appearance in early spermatids in many species. The dense chromatin on the nuclear membrane adjacent to the acrosome is generally characteristic of spermatids.

## THE ACROSOME

The acrosome is essentially a membrane-bounded granule packed with proteinaceous material which occurs as a cap overlying the anterior end of the nucleus. With the exception of teleost fish (*22, 88, 139*) and some insects (*104*), all spermatozoa that have been observed with the electron microscope possess acrosomes. Mammalian acrosomes contain hydrolytic enzymes (*81, 86, 90, 152*) believed to function in penetrating the cumulus and zona pellucida around the egg, thereby facilitating the sperm nucleus to reach the oocyte (*18, 19*). Acrosomes of many vertebrate and invertebrate species undergo precise morphological changes (*36a, 109a*) that are correlated with the entrace of the sperm nucleus into the egg. The size, shape, and internal structure of acrosomes varies greatly (Fig. 8). Some mammals, such as the guinea pig (*42, 50*) and golden mantled squirrel (Fig. 9), have large, elaborately shaped acrosomes. Other animals, for instance, some fish (*88*) and insects (*104*), have small, simple acrosomes.

**Fig. 8.** Acrosomes from a variety of animal species drawn to scale from electron micrographs showing the wide phylogenic diversity in acrosome size, shape, and internal morphology. (A) *Strongylocentrotus droebachiensis.* (B) *Arbacia punctulata.* (C) *Pseudocentrotus depressus.* (D) *Echinocardium cordatum.* (E) *Diadema setosum.* (F) *Asterias amurensis.* (G) *Asterina pectinifera.* (H) *Parastychopus californicus.* (I) *Gorgonocephalis eucnemis.* (J) *Comanthus japonicus.* (K) *Nereis japonicus.* (L) *Hydroides hexagonus.* (M) *Mytilis edulis.* (N) *Barnea candida.* (O) *Crassostrea virginica.* (P) *Acheta domestica.* (Q) *Procambarus clarkii.* (R) *Saccoglossus kowalevskii.* (S) *Rattus norvegicus.* (T) Rabbit. From C. B. Metz and A. Monroy, eds., "Fertilization," Vol. 1. Academic Press, New York, 1967. Courtesy of Dr. Jean C. Dan.

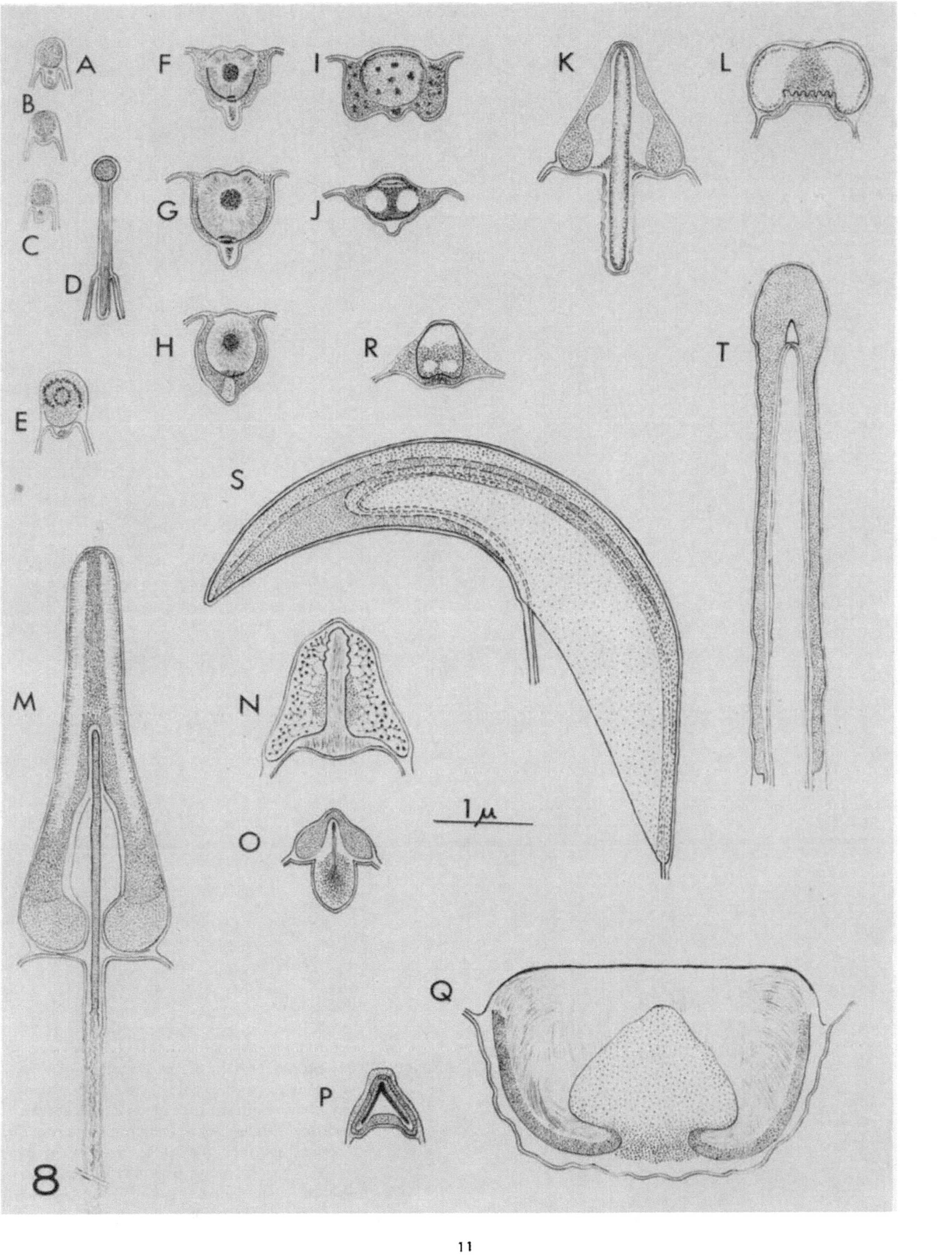
A
B
C
D
E
F
G
H
I
J
K
L
R
T
S
M
N
O
1μ
Q
P
8

**Fig. 9.** The anterior portion of the acrosome of three spermatozoa of the golden mantled ground squirrel in longitudinal section. In this species, acrosomes extend several microns anterior to the dense sperm nuclei at the bottom of the picture. The chromatin of squirrel sperm, unlike most mammalian species, never becomes completely compacted.

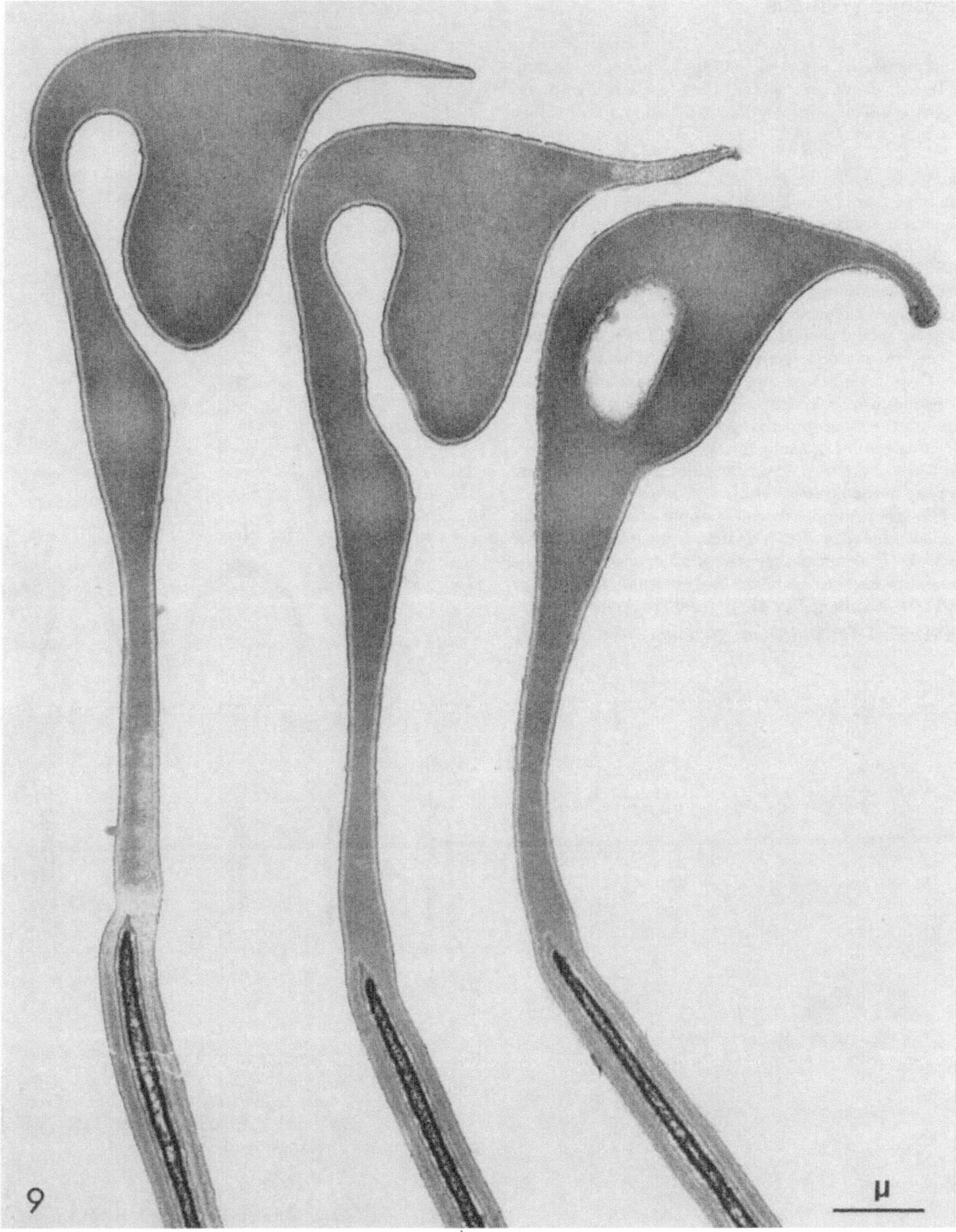
9
μ

## ACROSOME SYNTHESIS

Acrosomes are a specialized type of secretion granule or lysosome. When mature, they contain hydrolytic enzymes and remain membrane bound. As with other secretion granules (*14, 109*), the Golgi zone is intimately involved with their development. Acrosome formation generally begins with the appearance of a spherical, membrane-bound proacrosome granule on the concave surface of the Golgi complex between the Golgi and the nucleus, as is shown in Fig. 11. The location of the Golgi region and proacrosome granule defines the future anterior end of the sperm cell. This type of arrangement has been described in a wide range of animal species (*50, 72, 104, 133*). Disposition of the proacrosome granule on the concave side of the Golgi complex is analogous to secretion granule formation in most systems. However, in the exceptional case of the spur-throated grasshopper *Conocephalus*, the proacrosome granule forms on the convex side of the Golgi complex, i.e., on the side of the Golgi facing away from the nucleus (*104*).

The spherical proacrosome granule generally contains a small region of denser material oriented toward the nucleus. This remains on the nuclear side when the granule comes to lie over the nucleus (see Figs. 7 and 20). Acrosome formation generally occurs throughout spermiogenesis (Fig. 10).

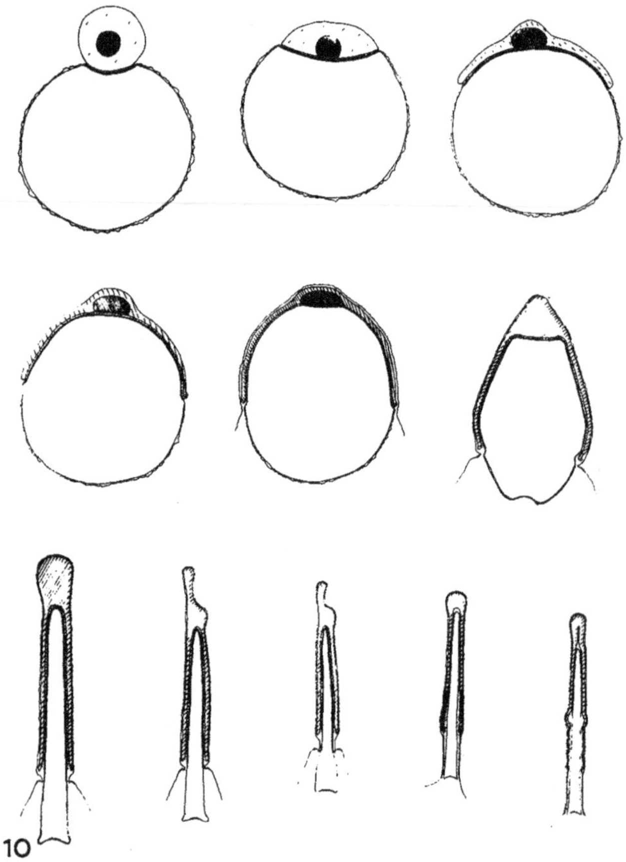
10

**Fig. 10.** The development of the acrosome during spermiogenesis in the rabbit. (Earliest stage, upper left; latest stage, lower right.) From *Z. Zellforsch. Mikrosk. Anat.* **115**, 553 (1971). Courtesy of Dr. L. Ploën.

**Fig. 11.** Acrosome formation in a spermatid of the stink bug *Euchistus*.

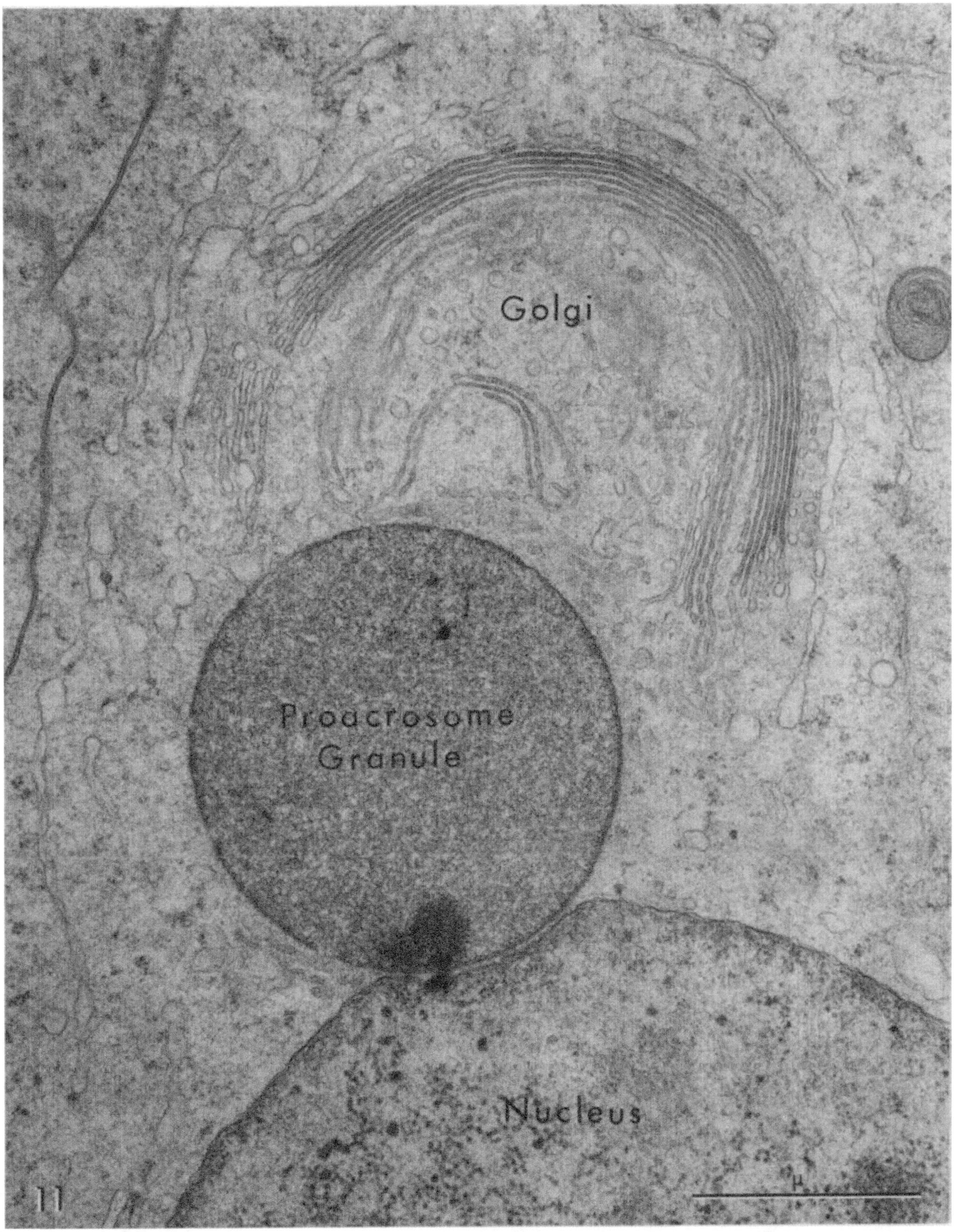
Golgi
Proacrosome
Granule
Nucleus
μ
11

## ACROSOME SHAPING

The forces that shape acrosomes into the precise and sometimes elaborate forms that they assume are not well understood. The manchette microtubules (see p. 22) are probably not involved in acrosome shaping since they begin posterior to the forming acrosome and are not associated with it. In mammals (Fig. 12), the anterior ends of the manchette microtubules are embedded in a ring of electron-dense, fibrillar material (the nuclear ring, see p. 30) just posterior to the acrosome. The forming acrosome is closely apposed to the Sertoli cell on one side and the nucleus on the other, associations that probably influence its shape. Internal forces, such as the precise packing of enzyme molecules within the acrosome, may also be partially responsible for the final shape (*108*). Mammalian acrosomes undergo changes in shape even after they are released from the testis (*20, 47, 50*). This may be correlated with the observation that the fertilizing capacity of spermatozoa gradually increases after they leave the testis as they pass through the epididymis (*17, 23, 56, 156*). Most invertebrate spermatids develop free in cysts without association with other cells. The acrosomes generally undergo precise changes in shape. This morphogenesis is presumably affected by molecular forces within the forming acrosome.

**Fig. 12.** Longitudinal section of elongating spermatid of the Chinese hamster.

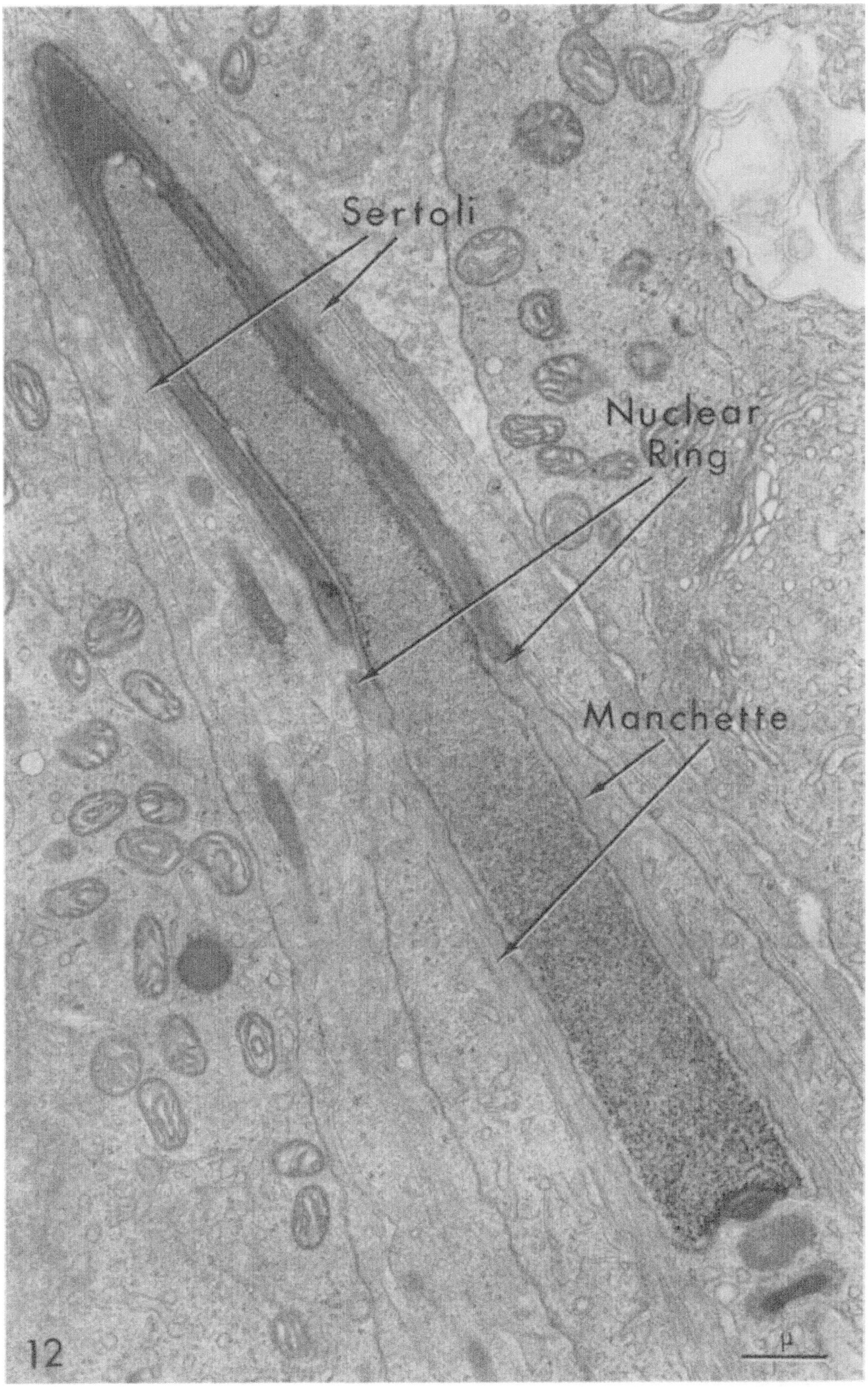
Sertoli
Nuclear
Ring
Manchette
12
1 μ

**Fig. 13.** The variety of shapes of sperm heads among some teleosts. A high degree of interspecies variability in sperm morphology is characteristic of most animal groups. (1) *Epiplatys senegalensis*. (2) *Clinus nuchipinnis*. (3) *Scorpaenia angolensis*. (4) *Chilomycturus antennaius*. (5) *Fistularia tobacaria*. (6) *Balistes forcipatus*. (7) *Upencus prayensis*. (8) *Cephalacanthus volitans*. (9) *Lebistes reticulatus*. (10) *Clarias senegalensis*. (11) *Hemichromis fasciatus*. (12) *Tráchinocephalus myops*. (13) *Pigusa triophthalmus*. (14) *Boops boops*. (15) *Antennarius senegalensis*. (16) *Ophioblennius atlanticus*. (17) *Parapristipoma octolineatum*. (18) *Zeus faber*. (19) *Fodiator aculus*. (20) *Blennius cristalus*. (21) *Papyracranus afer*. (22) *Sardinella aurita*. (23) *Periophthalmus papilio*. (24) *Galeoides decadactylus*. (25) *Ethmalosa fimbriata*. (26) *Blennius vanderveken*. (27) *Engraulis guineensis*. (28) *Lycodentis afer*. (29) *Tilapia nolotica*. From B. Baccetti, ed., "Comparative Spermatology." Academic Press, New York, 1970. Courtesy of Dr. X. Mattei.

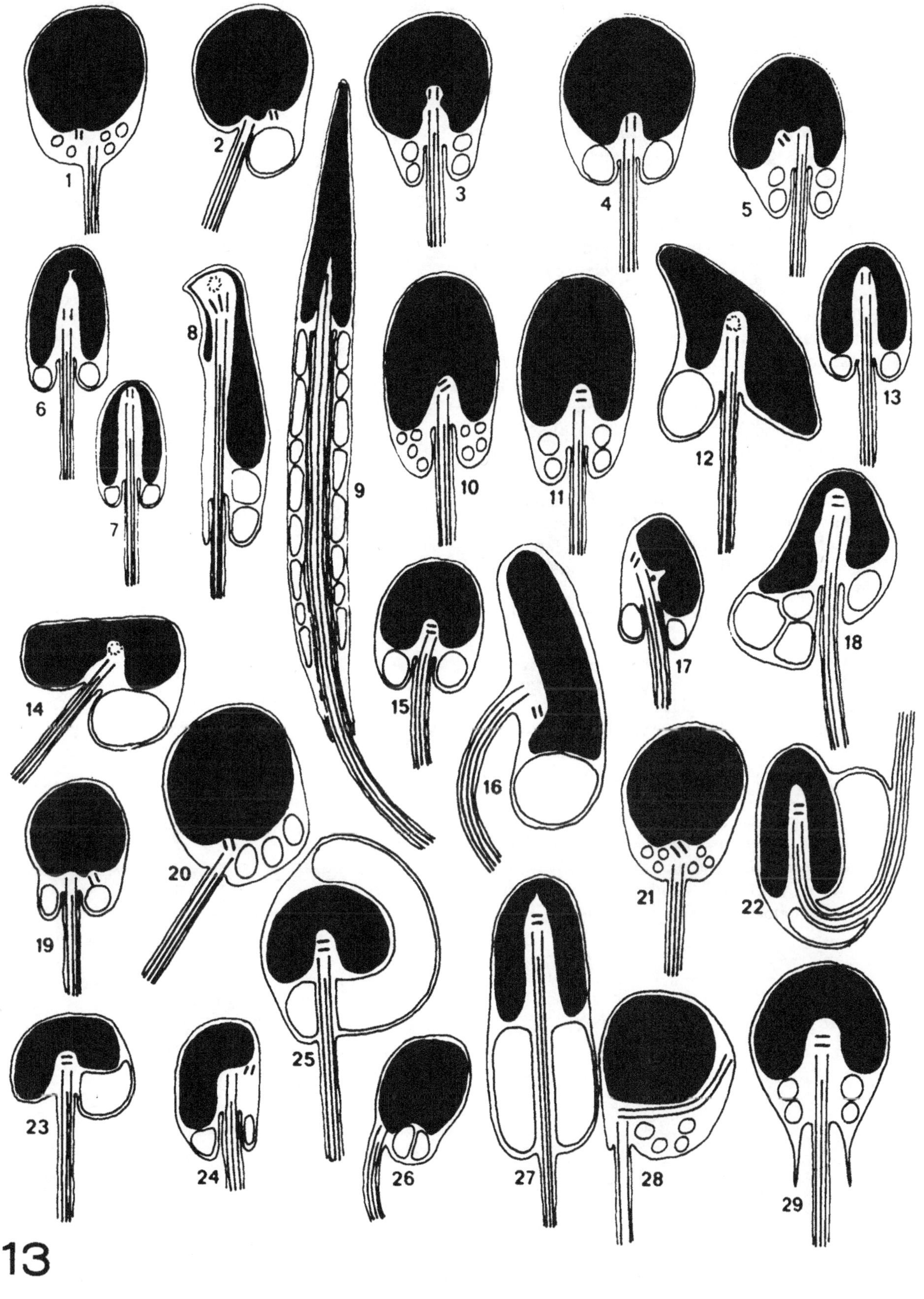


13

## THE NUCLEUS

Nuclei of mature spermatozoa are nearly all small, precisely shaped packages of highly compact inert chromatin. The dense appearance of the mouse sperm nucleus (Fig. 14) is typical. The compacted, elongate nuclear shape presumably is consistent with efficient motility.

During spermiogenesis, chromatin of many species has been observed to display very regular, or even geometric, patterns (*104*). In some species, chromatin condensation occurs in specific areas of the nucleus in a specific sequence. There appears to be no pattern of chromatin condensation general to all species. In some species, chromatin condenses around the periphery of the nucleus, whereas, in other, condensation appears to begin in the anterior of the spermatid nucleus. Spermatid nuclei of some species display longitudinally oriented lamellae, whereas, in spermatids of other species, chromatin appears more fibrous. In mature spermatozoa of mammals, freeze etching studies reveal chromatin apparently arranged in stacked lamellae lying in the plane of the flattened sperm head (*78, 110*). Several lines of experimental evidence suggest that in insects (*69, 141*) and amphibians (*84, 84a*) chromosomes may be packed in a precise sequential order within sperm nuclei.

Chromatin of sperm nuclei is chemically as well as structurally altered during spermiogenesis. Arginine-rich histones or protamines replace lysine-rich histones of early spermatids (*1, 24, 25, 38, 52, 73, 112*). Protein synthesis, except in very young spermatids, appears to be limited primarily to the synthesis of argine-rich nuclear proteins (*92*). RNA synthesis occurs only during early spermiogenesis, and, in most systems which have been studied, ceases as soon as nuclear condensation and elongation begins (*66, 92, 94, 144*). However, in the earthworm, nuclear RNA synthesis is reported to continue until late in spermiogenesis (*3, 4*).

**Fig. 14.** Longitudinal section of mouse sperm head. Small vesicles coat a region of the plasma membrane overlying the acrosome.

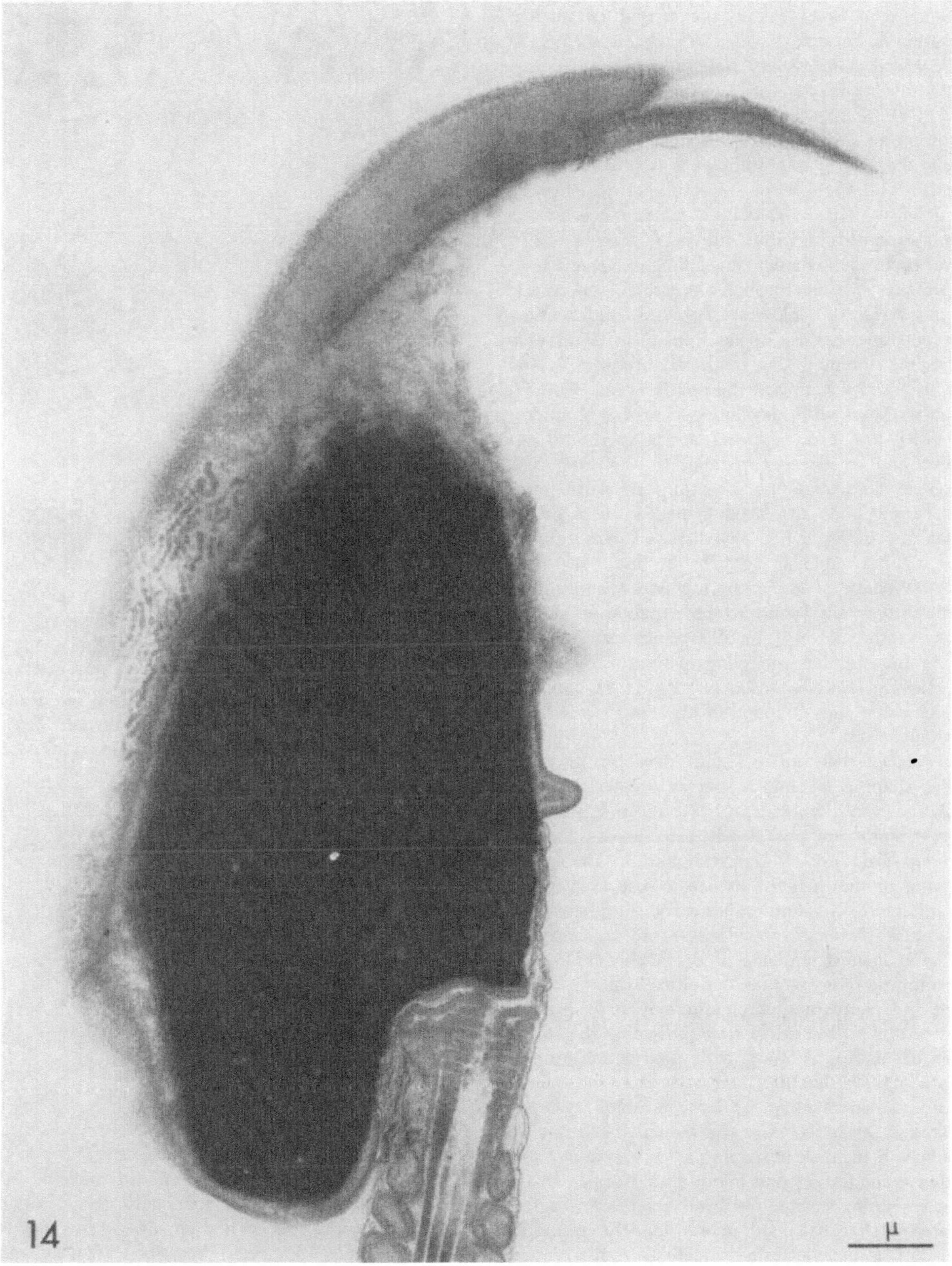
14
μ

## THE LONGITUDINAL MANCHETTE

During spermatid development, the nucleus is surrounded by a microtubular array. Microtubles encircle the spermatid nucleus during the period of nuclear condensation in flatworms (*33, 138*), annelids (*6, 7, 9, 28, 53*), insects (*74, 75, 99, 104, 137, 147, 154*), reptiles (*26*), birds (*87, 89, 155*), and mammals (Fig. 16) (*26, 30, 102*). They extend from the posterior margin of the acrosome to several microns behind the nucleus. In mammals, the ring of microtubules is referred to as the manchette. The number of microtubules constituting the manchette varies considerably among species but is strikingly consistent within the same species (117).

Several investigators analyzing different systems have suggested that the microtubules encircling spermatid nuclei play a role in nuclear shaping and condensation. In dragonfly spermatids, nuclei appear in transverse section as multipointed stars with microtubules in the furrows only and not around the points (Fig. 15) (*74, 76*). Rooster spermatids develop two arrays of microtubules (*89*). One array appears late in spermatid development and is directed parallel to the long axis of the spermatid nuclei; the other appears earlier and contains microtubules displaced in helices. It is postulated that the sliding of helically disposed microtubules over one another might account for the final very elongate, narrow shape of the sperm nucleus. Microtubules in other systems are found to be capable of sliding over one another (*134*). In all species studied, the manchette microtubules encircling spermatid nuclei disappear after nuclear condensation is complete, supporting the idea that they function in nuclear shaping during spermiogenesis.

Factors other than microtubules may be involved in nuclear shaping in certain species. In at least one mammal, the woolly opossum (*102*), the microtubules of the manchette are not closely associated with the nucleus and are directed perpendicular to the plane of flattening of the nucleus. In one teleost (*139*), the nuclear perimeter remains spherical during spermiogenesis while chromatin condenses into a crescent-shaped band against one side of the nucleus. The nuclear membrane later overlies the chromatin. Thus the shape of the condensed chromatin may be the main factor determining the nuclear shape in this fish. Such instances of nuclear shaping with no relation to the manchette suggest that the manchette may not be involved in nuclear shaping, at least in some systems. Some workers even feel that the manchette is not directly involved in nuclear shaping in any system since the evidence suggesting that microtubules are involved in nuclear shaping is based entirely on morphological correlation. It has been suggested that the principle function of the manchette may relate to redistribution of cytoplasm during spermiogenesis (*45*).

15

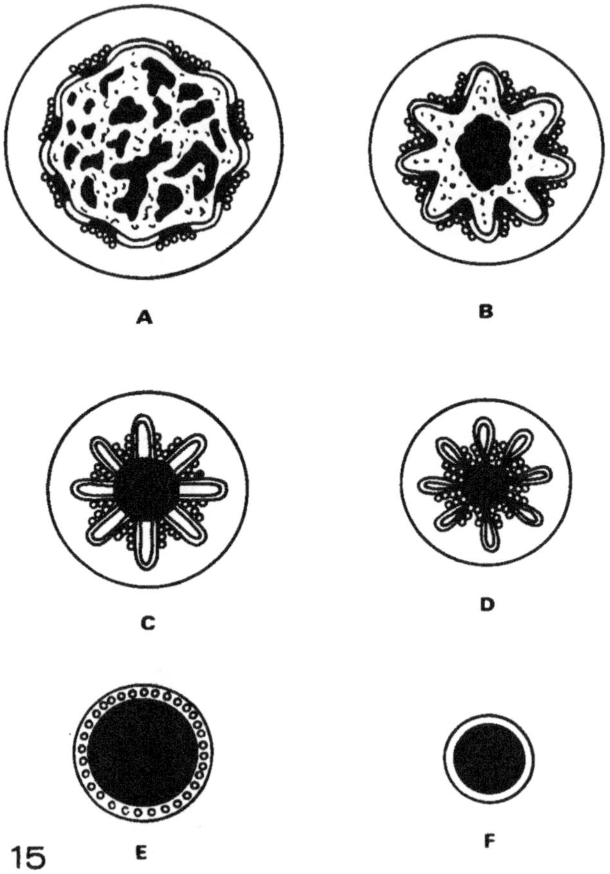


**Fig. 15.** The morphological relationship between microtubules and nucleus during nuclear condensation and elongation in dragonfly spermatids. From *J. Ultrastruct. Res.* **16**, 293 (1966). Courtesy of Dr. R. G. Kessel.

**Fig. 16.** Chinese hamster spermatid nucleus. Microtubules of the manchette are frequently evenly spaced and in rows. Arms can sometimes be seen between adjacent microtubules.

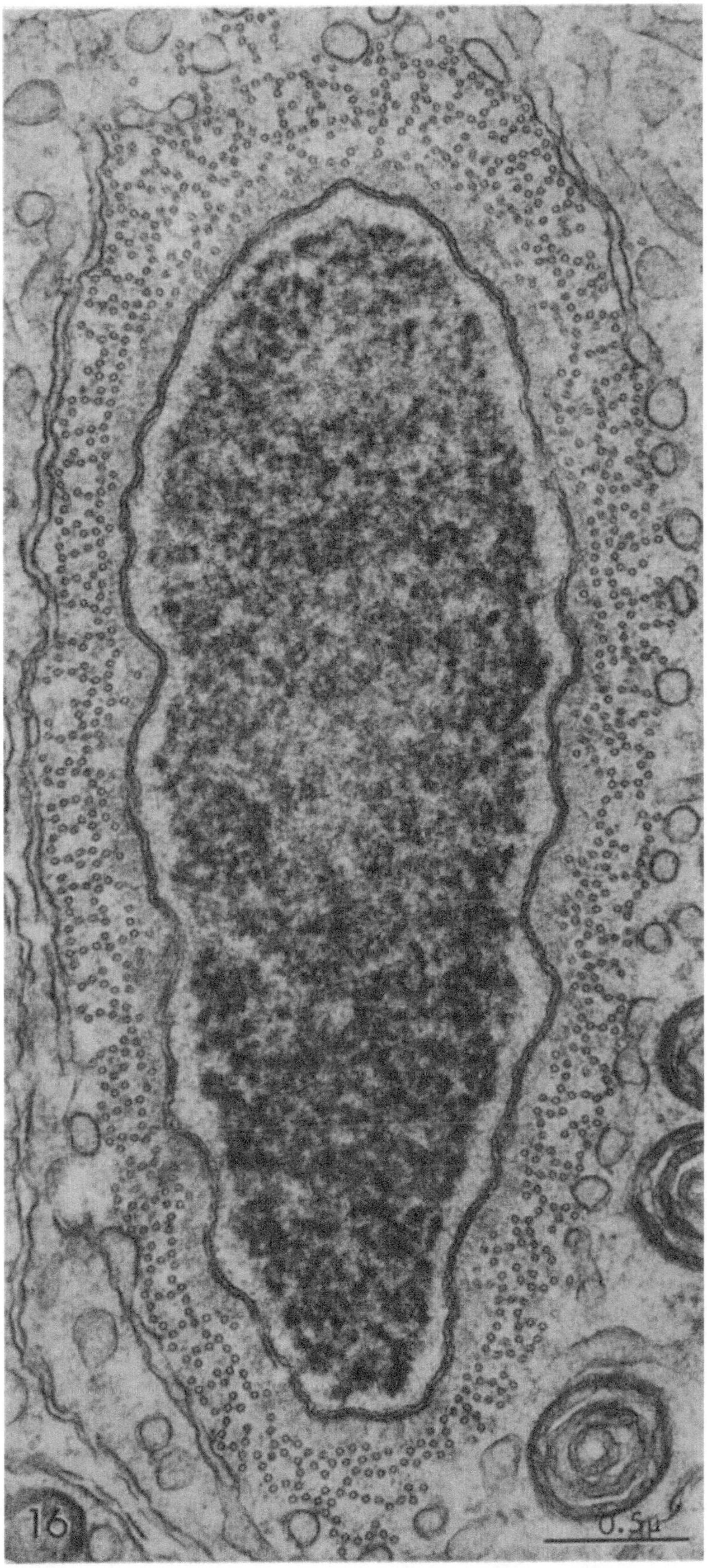
16
0.5μ

## THE CIRCULAR MANCHETTE

Although the manchette is generally composed of microtubules disposed parallel to the elongating spermatid (see p. 22), a manchette of microtubules disposed perpendicular to the long axis of the cell nucleus occurs in young spermatids of chickens (89), in three American marsupials (*114*), and in several lizards (*34*). For example, in Fig. 18 of a young spermatid of the banded gecko, the manchette microtubules are seen to encircle the nucleus. Analysis of serially sectioned spermatids suggests that the circular manchette in the rooster consists of two helically arrayed tubules rather than concentric circles (89). The sliding of juxtaposed microtubules of the circular manchette may be responsible for the elongation and condensation of nuclear contents. If this is generally true, one would expect equal numbers of tubular profiles to appear on either side of the nucleus in longitudinal sections. In the circular manchette of early opossum spermatids, there are significant differences between the number of microtubule profiles on each side of longitudinally cut spermatid nuclei (*114*), suggesting that this circular manchette consists of numerous discontinuous microtubules rather than one or two microtubules spiraling the length of the nucleus. In all species where it has been seen, the circular manchette is a transient structure occurring only in very young sperm. It is later replaced by a typical parallel manchette. The two arrangements can be compared in the micrographs of an immature (Fig. 18) and more mature (Fig. 19) spermatid of the banded gecko. In Fig. 19, the microtubules of the parallel manchette tend to be evenly spaced in rows and are sometimes connected by small arms. No transition stages have yet been described so it is not clear whether the microtubules of the circular manchette are reutilized in the formation of the longitudinal manchette. Arms connecting microtubules are common to all manchettes. They have also been described in systems in which microtubules appear to slide on one another, such as the feeding organelles of protozoans (*142, 143*) and protozoan axostyles (*63*), and arms are also present in spindles (where microtubules may possibly slide on one another) (*67*). Some workers feel that this correlation supports the supposition that microtubules of the manchette are involved in some sort of active movement.

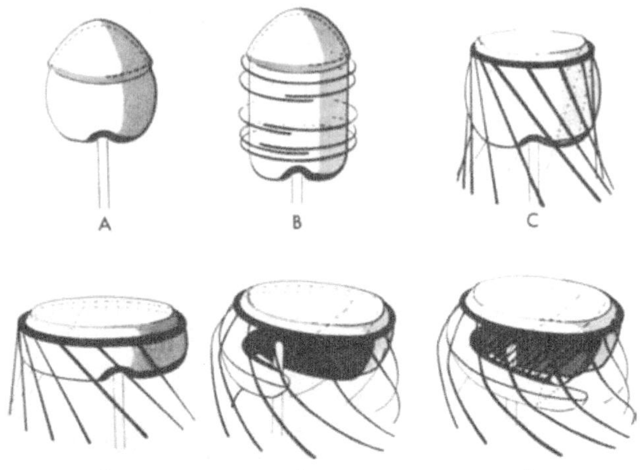


17

**Fig. 17.** Stages of spermiogenesis in the opossums, *Marmosa mitus* and *Didelphus virginiana.* (A) Early stage of spermiogenesis with the anterior–posterior axis established by the placement of the acrosomal cap and implantation fossa. (B) Ellipsoidal nucleus encased in a cylindrical sheet of microtubules. (C, D) Spermatid showing anterior migration of the posterior margins of the nucleus. Obliquely arranged microtubules encase the nucleus and are associated with a prominent nuclear ring. (E) Spermatid in which a posterior protrusion of the nucleus indicates the initial formation of the minor arm of the U-shaped head. (F) Spermatid in which nuclear condensation and shaping has been completed. Adjacent microtubules are associated laterally by dense matrix. From *J. Ultrastruct. Res.* **40**, 498 (1972). Courtesy of Dr. J. B. Rattner.

**Fig. 18.** Transverse section of early spermatid of the banded gecko, *Coleonyx variegatus brevis,* showing the transitional circular manchette.

**Fig. 19.** Transverse section of late spermatid of *Coleonyx variegatus brevis* showing longitudinal manchette.

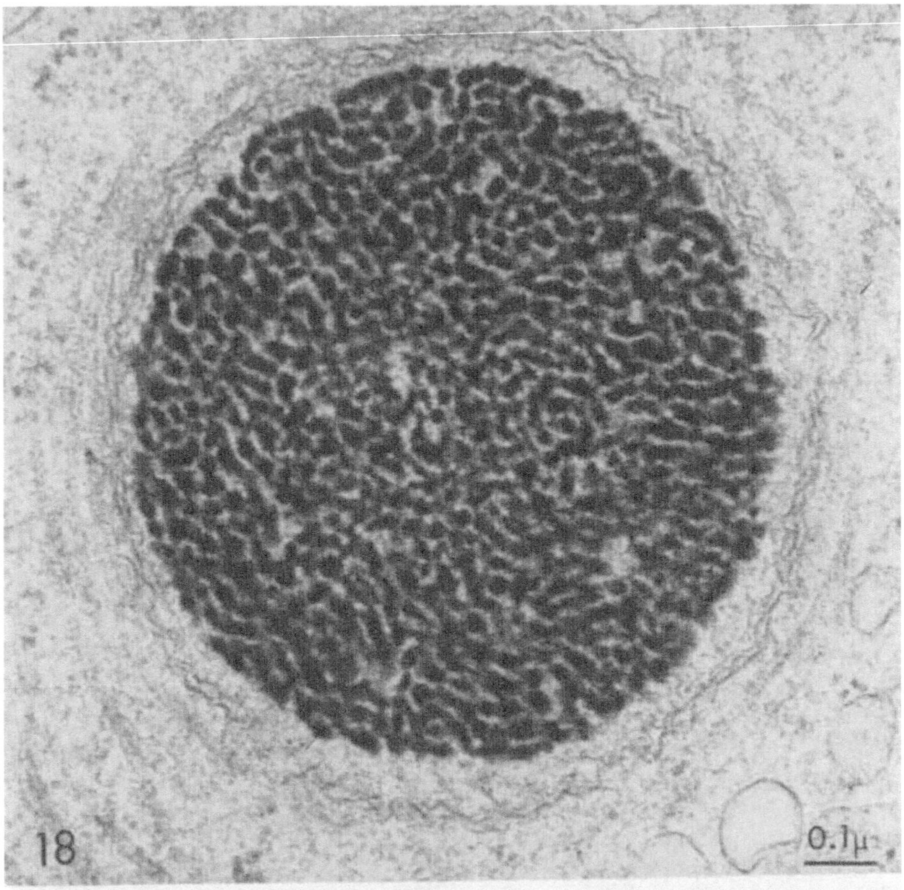
18
0.1μ

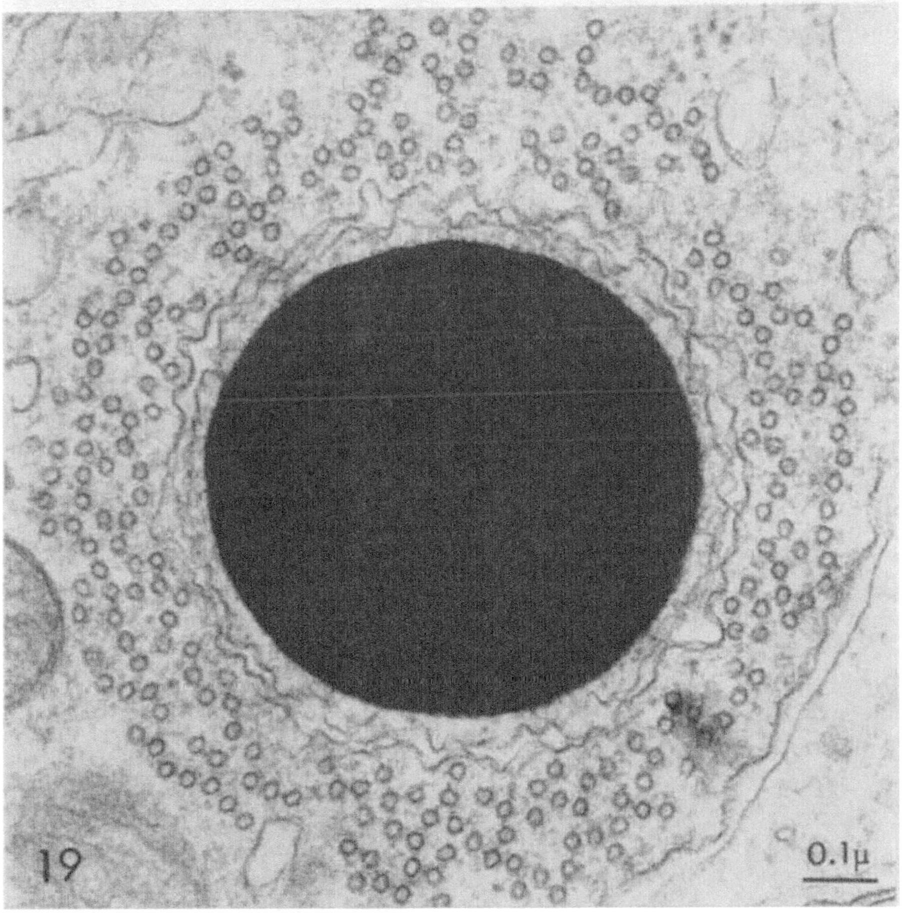
19
0.1μ

## FLAGELLAR GROWTH

Very young spermatids generally have one pair of centrioles which are initially located just below the plasma membrane. Early in spermiogenesis, the flagellum begins to extend from one centriole. This centriole is referred to as the distal centriole. The centriole pair with flagellar tubules extending from it migrates later from just below the plasma membrane to a new position near the nucleus. Dense material begins to accumulate around the centrioles during the centriolar migration and eventually becomes very extensive. It appears to function in attaching the base of the flagellum to the nucleus. In mammals, it increases notably in amount and becomes the columns of the connecting piece (Fig. 23). In insects, similar material is termed the centriole adjunct (*32*). Material associated with centrioles has also been described in spermatids of birds (*96*), fish (*88*), amphibians (*92, 148*), reptiles (*55*), and earthworms (*7*). Figure 20 illustrates some of the changes that occur as the centriole pair migrates.

A local involution of the nuclear membrane forms where the proximal centriole is juxtaposed to the nucleus. This involution, termed the implantation fossa in mammals (*49*), has a regular thickened appearance due to the chromatin lining. A similar relationship between the nucleus and centriole is seen in almost all other groups (Fig. 41).

During these early stages of flagellar growth, the plasma membrane becomes attached to a ring of dense material, the annulus or "ring centriole," which behaves as if it were anchored near the anterior end of the flagellum. As the flagellum elongates and the centrioles migrate, the plasma membrane maintains its association with the annulus, becoming involuted around the flagellum in order to do so. The annulus later moves posteriorly, carrying the plasma membrane with it (see Fig. 43). A similar process occurs in insects, birds, and reptiles.

**Fig. 20.** Young spermatid of the rabbit. In this micrograph, a cytoplasmic bridge (arrow) can be seen connecting the spermatid to another spermatid. The forming acrosome is slightly more mature than the acrosome of the rabbit spermatid shown in Fig. 7. Multivesicular bodies, as seen here, are common in young spermatids.

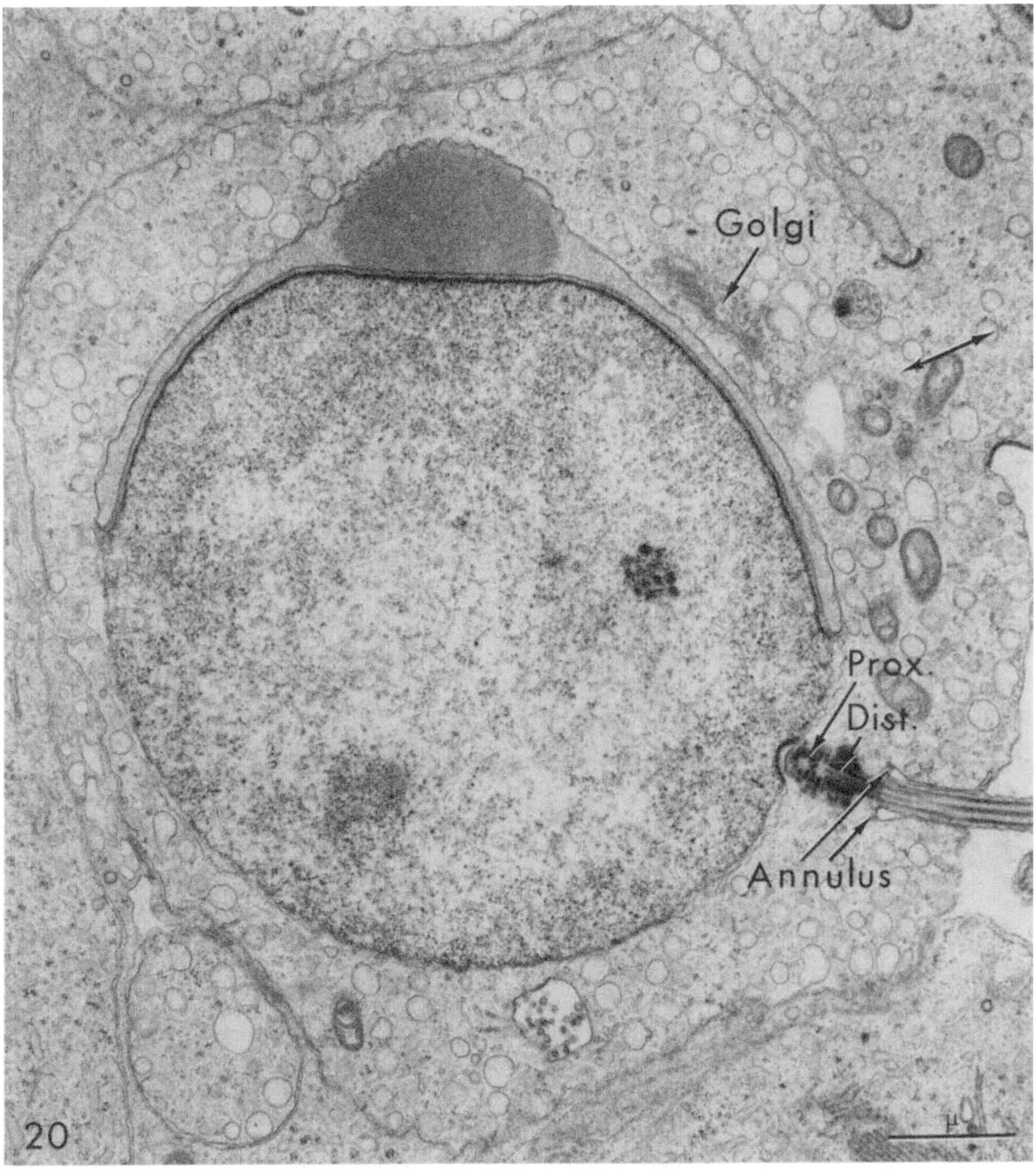
Golgi
Prox.
Dist.
Annulus
20
μ

## THE NECK REGION OF MATURE SPERMATOZOA

The region of a spermatozoon just posterior to the sperm head is termed the neck. The neck region shown in this longitudinal section of a golden mantled ground squirrel spermatozoon (Fig. 22) is particularly elaborate. A great deal of structured dense material forms in relation to the proximal centriole where it closely apposes the posterior edge of the nucleus. The distal centriole, which served as a basal body for the flagellum as it formed, characteristically disappears during mammalian spermiogenesis. In squirrels, the flagellar tubules terminate more than a micron posterior to the sperm head. Generally the distance is much less in other species. In the region of the midpiece (posterior to the neck region), the section transects the extensive network of thin satellite fibers found between the dense fibers in sperm of several mammals (*48, 51*). The satellite fibers are particularly prominent in squirrel spermatozoa. Spermatozoa of many invertebrate species also have intricate structures in the neck region (Fig. 21).

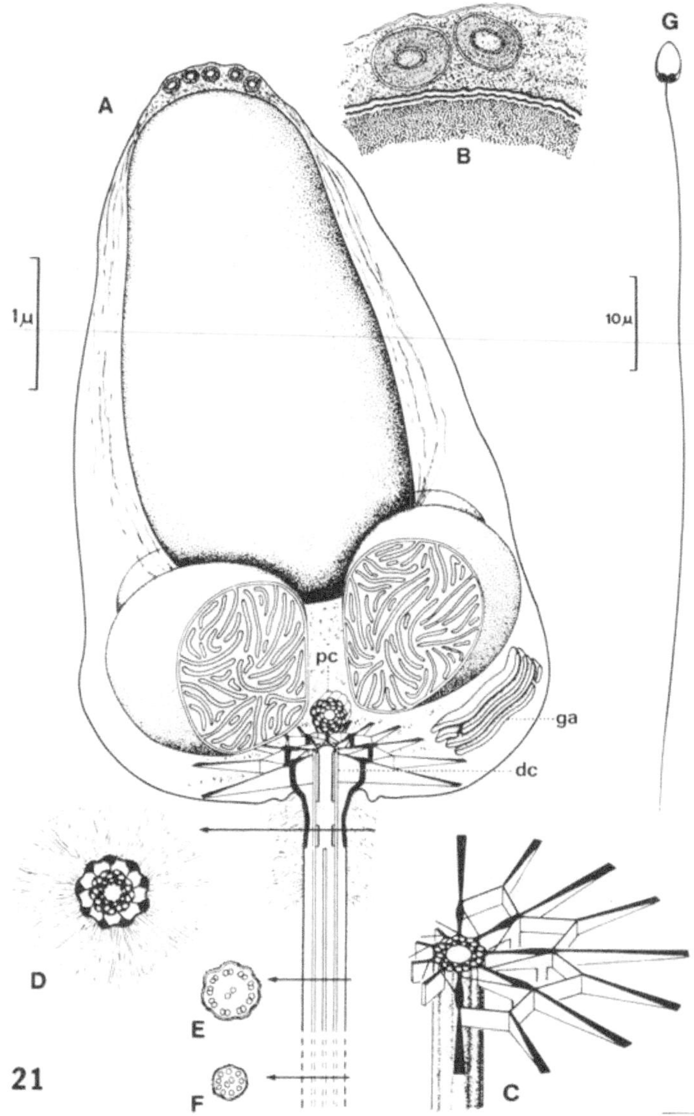


21

**Fig. 21.** A mature spermatozoon of the jellyfish *Naustihoe*. In this species, an elaborate structure is associated with the distal centriole. This structure presumably serves to anchor the flagellum in place. As is often the case in invertebrate species that produce spermatozoa that swim through seawater, this sperm is very simple. The 9 + 2 flagellum has no accessory structures, and a small number of mitochondria, generally four, are located directly behind the head in a radial array. (A) Semischematic diagram of the spermatozoon of the jellyfish *Nausithoe*. The short conical head consists of the nucleus, a relatively prominent cytoplasm containing membranes, and in the place of the acrosome a layer containing vesicles. In the midpiece, four large mitochondria are seen as well as a Golgi apparatus (ga), the proximal centriole (pc), and the distal centriole (dc) suspended in its anchoring fiber apparatus. The proximal part of the tail has a peculiar hairy coat consisting of slender fibrils. Only the proximal 1.5% of the tail is included in the diagram. (B) The anterior end of the sperm head at a higher magnification. The complex appearance of the vesicles is seen. (C) The anchoring fiber apparatus and the distal centriole. (D) Cross section of the tail at the level of the hairy coat. (E) Cross section of the tail at the level of its main portion. The surface is covered with short projections which may be a glycocalyx. (F) Cross section of the tail at the level of the end piece. Some single filaments at this level have a triangular outline. (G) Diagram of the living spermatozoon drawn from light microscopical observations. From *J. Ultrastruct. Res.* **37**, 195 (1971). Courtesy of Drs. Afzelius and Franzen.

**Fig. 22.** Longitudinal section of the neck region of a mature spermatozoon of the golden mantled ground squirrel. The proximal centriole is embedded in dense material. The distal centriole has degenerated. The neck region of squirrel spermatozoa is particularly broad and long.

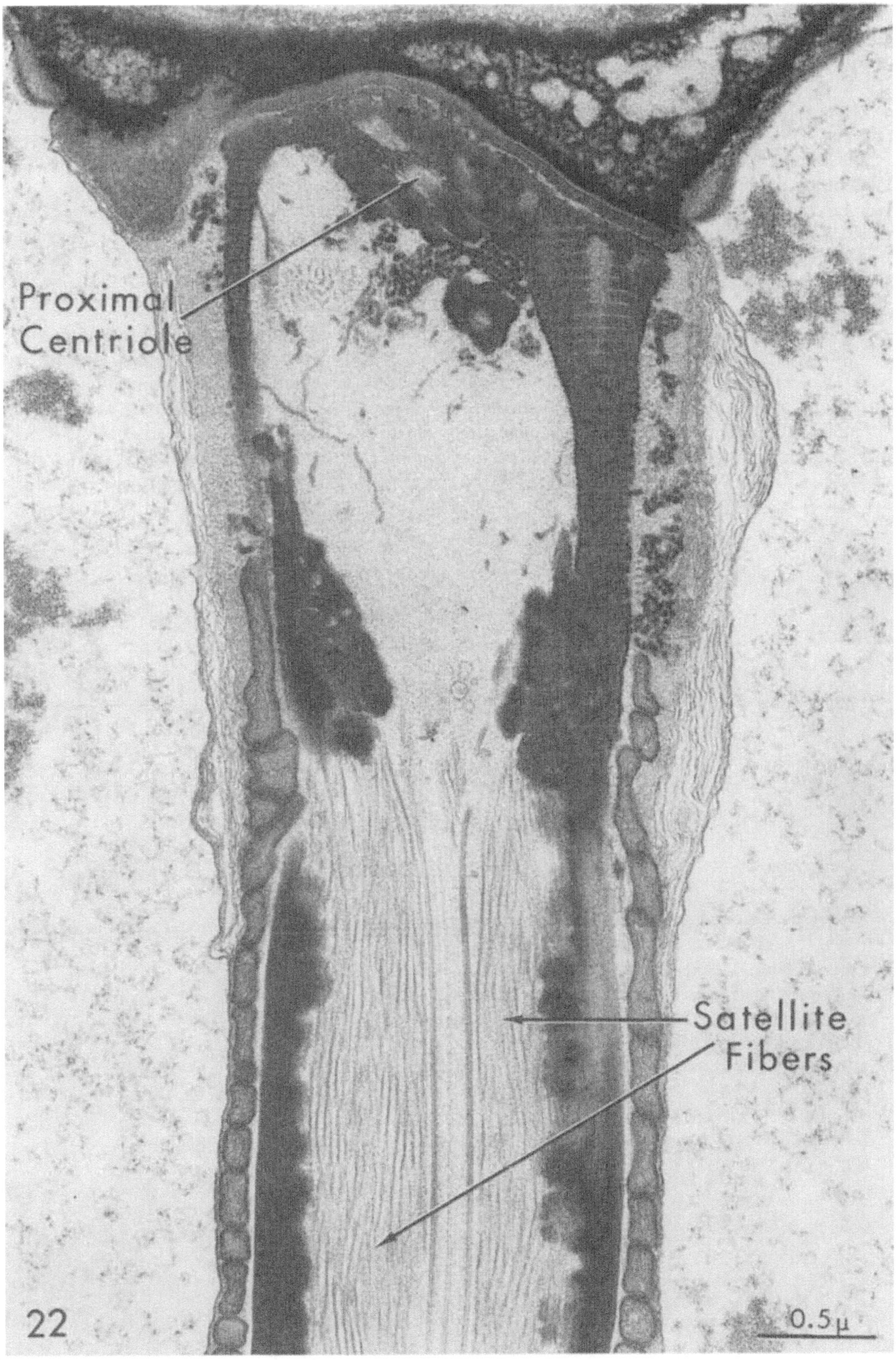
Proximal Centriole
Satellite Fibers
22
0.5μ

## DEVELOPMENT OF THE NECK REGION

The complicated events that occur during the morphogenesis of the neck have been followed in mammals (Fig. 23) (*49*), insects (*32*), amphibians (*82*), and the earthworm (*7*). Figure 24 shows a developing neck of the Chinese hamster sperm. Striated columns of the connecting piece are the most prominent features. The nine columns originally form in realtion to the distal centriole (basal centriole), which disappears later during spermiogenesis. They have a major period of 665 Å. Each dense band is 520 Å long and is subdivided into ten evenly spaced narrow bands. This periodicity is similar to that observed in ciliary rootlets, which are also centriole-related structures thought to be involved in anchoring basal bodies. The nuclear membrane in late spermatids characteristically extends posteriorly into the neck region. In some species, the nuclear membrane extends far beyond the neck region into the midpiece (*116*). The posterior portion of the nucleus around the proximal centriole assumes a precise shape. Dense material accumulates on both sides of the nuclear membrane in this region, which is termed the articular fossa (Fig. 1). Observations of living mammalian spermatozoa show that the neck region is stiff and that there is no articulation between flagellum and head in the neck region (*62, 108, 127*).

The nuclear ring of dense material in which the manchette microtubules are embedded moves posteriorly during the course of spermiogenesis. In Fig. 24, it is near to the neck region. The nuclear ring is generally associated with the posterior margin of the acrosome which spreads posteriorly over the nucleus in conjunction with the movement of the ring (Fig. 43). The ring reaches the posterior end of the nucleus at about the time that nuclear condensation is completed; then both the ring and its microtubules disappear. In Fig. 24, the large annulus with attachment to the cell membrane has yet to migrate posteriorly and is still at the posterior end of the neck.

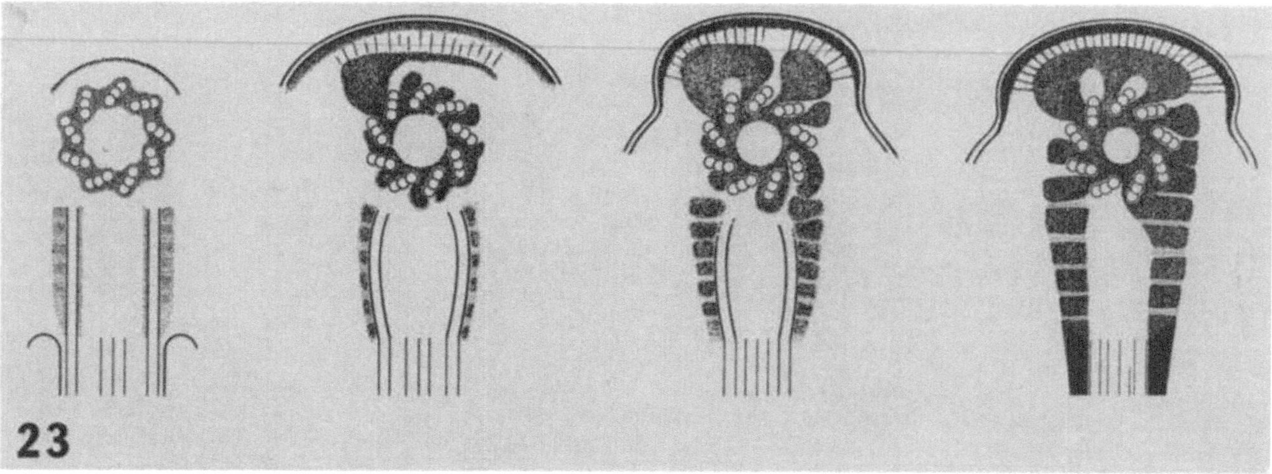

**Fig. 23.** Stages in the development of the connecting piece (most immature stage at left). The dense material that composes the columns of the connecting piece and indention in the nucleus arises in close association with the distal and proximal centrioles. From B. Baccetti, ed., "Comparative Spermatology." Academic Press, New York, 1970. Courtesy of Dr. D. W. Fawcett.

**Fig. 24.** Neck region of a Chinese hamster spermatid.

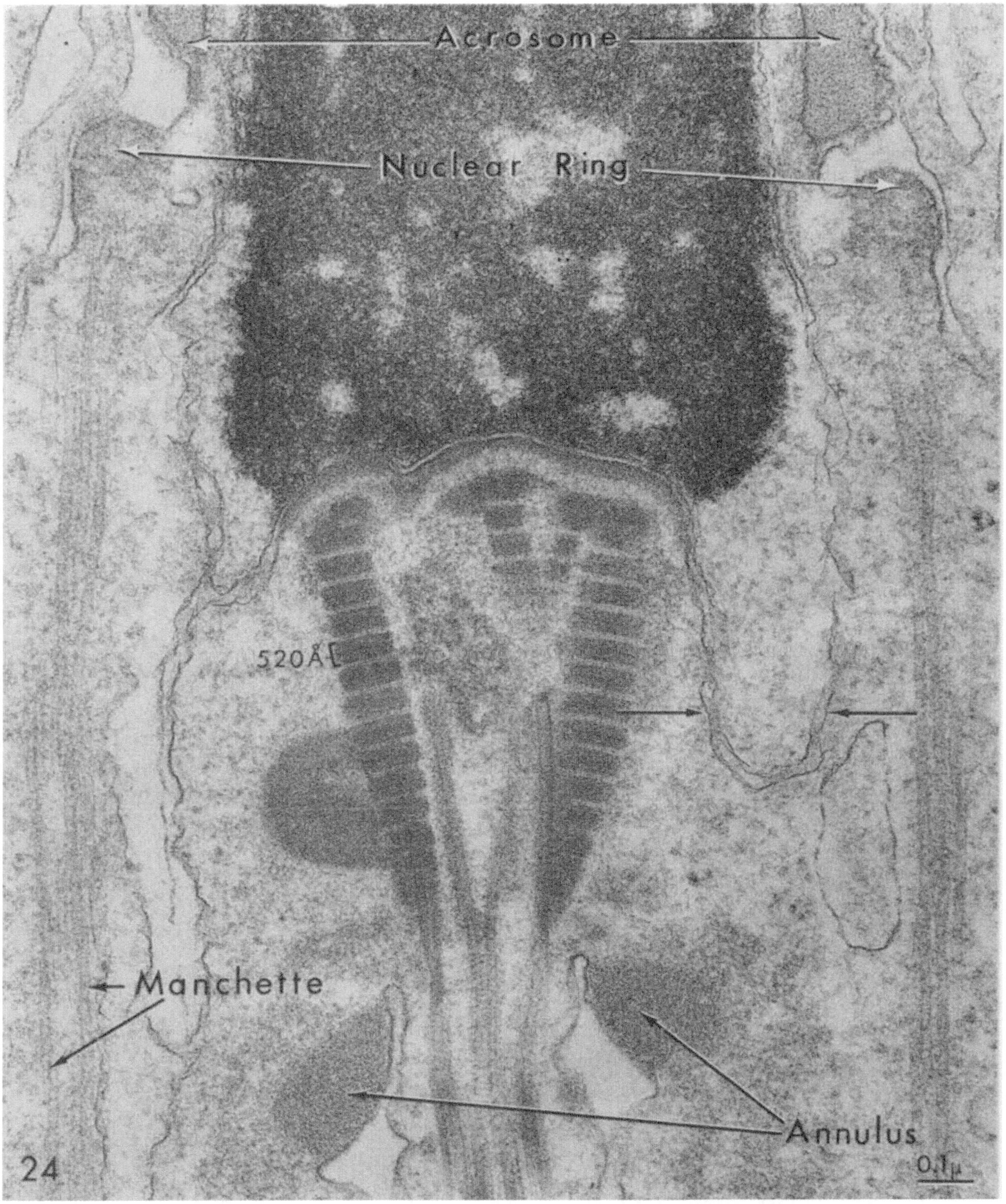
Acrosome
Nuclear Ring
520Å
Manchette
Annulus
24
0.1μ

## MIDPIECE

In mammals and other vertebrates, there is a region between the neck and the annulus of a spermatozoon termed the midpiece or middle piece. In the light microscope, it appears thicker than more posterior regions because the flagellum there is surrounded by mitochondria. In the transverse section of mammalian sperm, the midpiece is seen to contain the characteristic 9 + 2 flagellum, nine dense or coarse fibers, and the sausage-shaped mitochondria (Fig. 26). As in other 9 + 2 flagella, one of the two tubules of each of the nine doublets (sometimes called outer fibers) is slightly smaller than the other and generally has a more electron-dense center. Two arms can often be seen extending from this tubule in the direction of the adjacent doublet. This tubule is termed subfiber A and the other member of the doublet is called subfiber B (Fig. 29). The nine doublets are numbered in the following manner: If a line is drawn directly between the two tubules of the central pair and perpendicular to a line that passes through the center of each of them, the line passes through one of the doublets. This doublet is numbered number one. If the arms of subfiber A of the doublet are directed clockwise (i.e., the flagellum is viewed from base to tip), the next doublet in a clockwise direction is number two, and the next number three, and so on. Dense fibers are given the number of the doublet with which they are associated. (See *57* for further explanation.)

Dense fibers have been described in birds (*45, 95*) and reptiles (*55, 65*) as well as mammals. Insects have accessory tubules in positions analogous to the position of dense fibers (Fig. 29). Sperm flagella of some animals have no accessory elements at all. The size and shape of dense fibers is characteristic for a species. Some, like the Chinese hamster, have large dense fibers; others, such as man (*98*), rabbit (*157*), and opossum (*104*), have small, dense fibers. Although the size and shape of dense fibers vary among species, there are similarities; the dense fibers associated with flagellar doublets one, five, and six are larger than the others at most levels, and dense fibers numbers three and eight terminate most anteriorly in all species. Viewed in transverse section, one can usually distinguish an inner and a cortical region of each dense fiber. The inner and cortical regions display markedly different staining properties under certain conditions (*59, 61*). Motile sperm with large dense fibers have a much stiffer beat than do motile sperm with small dense fibers (*108*). The dense fibers may, therefore, be stiff elastic structures. However, ATPase activity has been localized in the cortical region of dense fibers with EM cytochemical techniques (*60, 95*), suggesting that dense fibers may have a contractile function as well.

The dense fibers arise after the formation of the 9 + 2 tubules of the flagellum. Each dense fiber forms in association with a doublet. They appear first as thin rods that extend the full length of the flagellum and gradually increase in diameter to assume their distinctive morphologies. Dense fibers are presumably composed of a highly order array of molecules since they are birefringent and display a periodic oblique banding in thin section electron micrographs (Fig. 25).

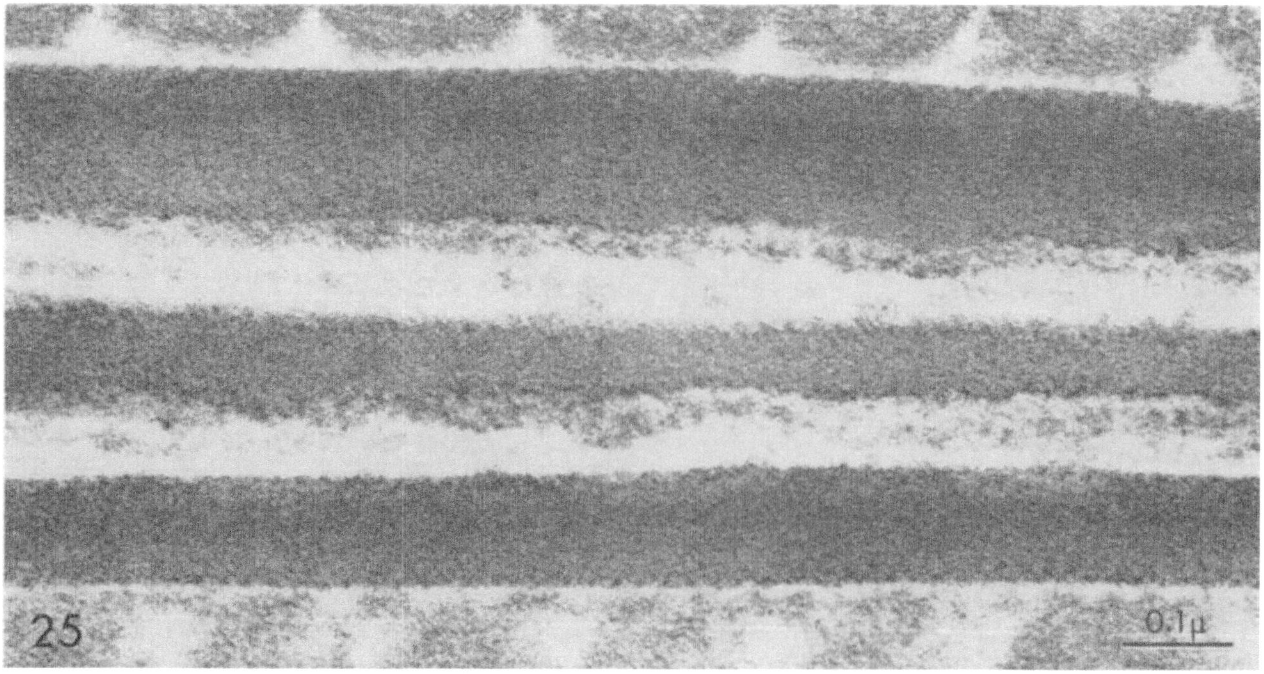


**Fig. 25.** Longitudinal section of dense fibers in a rat spermatozoon. Dense fibers show periodic structure at an angle of about 60° to the long axis of the sperm.

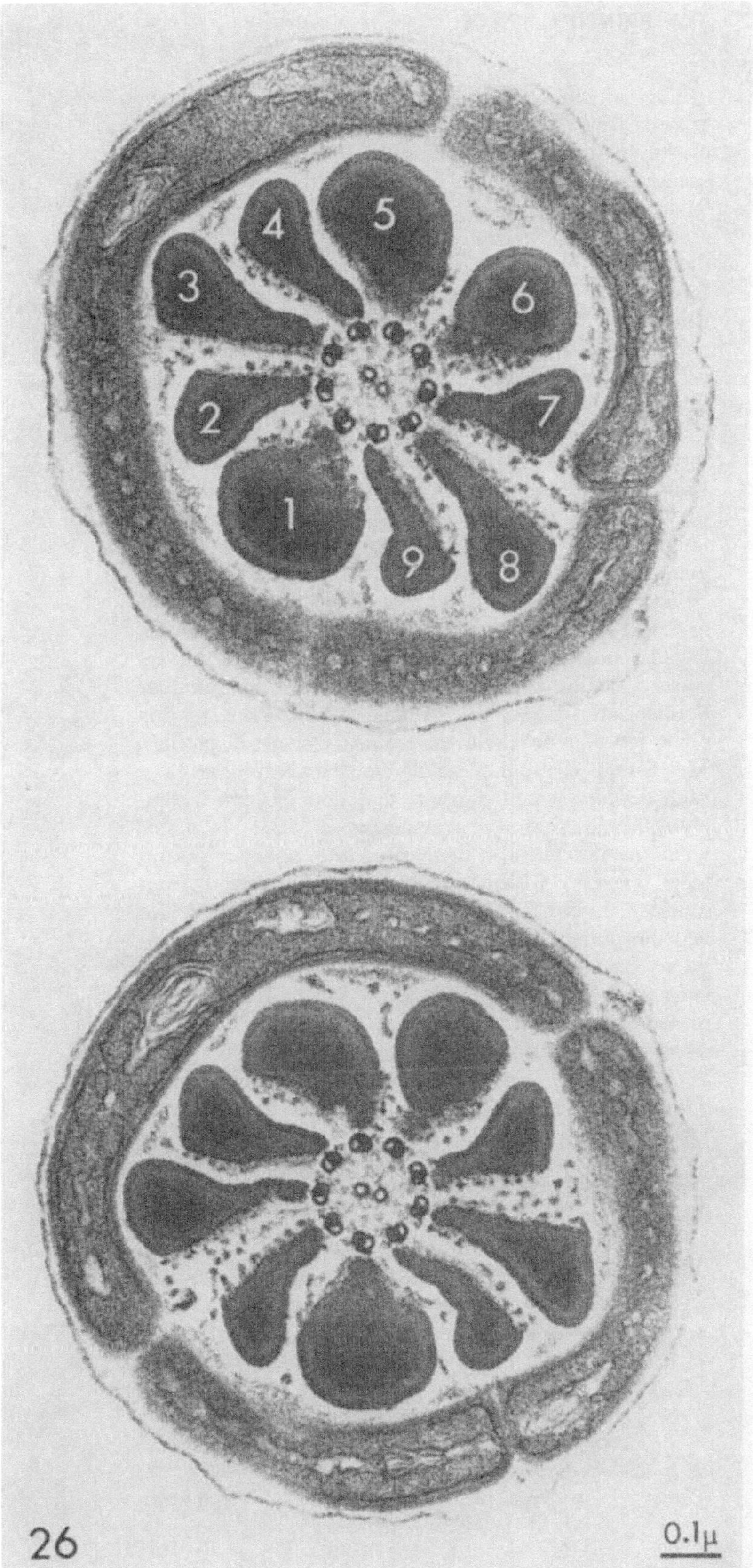


**Fig. 26.** Transverse section through the midpiece of two Chinese hamster spermatozoa.

## THE PRINCIPAL PIECE

The principal piece of a mammalian spermatozoon extends from the annulus posteriorly nearly to the end of the sperm cell. It contains the 9 + 2 flagellum and seven of the nine dense fibers from the midpiece, all enclosed by a prominent fibrous sheath. The fibrous sheath is composed of a cylinder of material beneath the plasma membrane. The cylinder is thickened to form two prominent longitudinal columns which make contact with doublets three and eight on either side of the flagellum (Fig. 28). These doublets have associated dense fibers in the midpiece which do not extend into the principal piece. The dense fibers and fibrous sheath gradually taper along the principal piece and terminate just anterior to the end of the spermatozoon. Cells cut posterior to this through the end piece display just the 9 + 2 flagellum (Fig. 28). Viewed in longitudinal section, the fibrous sheath appears segmented (Fig. 27).

Flagellar doublet tubules terminate in a precise manner at the end of the sperm tail. Near the tip of insect flagella, posterior to the termination point of the accessory tubules and central pair, subfiber A and subfiber B dissociate. Transverse sections cut posterior to this point reveal nine circles and nine "C-shaped" profiles. The C-shaped profiles are derived from subfiber B of each doublet which suggests that doublets are incomplete tubules sharing a common wall (*104*). In mammals, the doublet also dissociates into nine circles and nine C's, the C-shaped tubule being derived from subfiber B, but in the most posterior regions of the flagellum, the C-shaped profiles close and become complete circular tubules (*105*). C-Shaped tubules have been described in a number of microtubular systems in which they are believed to represent a stage in microtubule assembly or disassembly (*21, 36, 132*). In spermatozoa, however, they appear to represent a permanent feature of a short region of the flagellum.

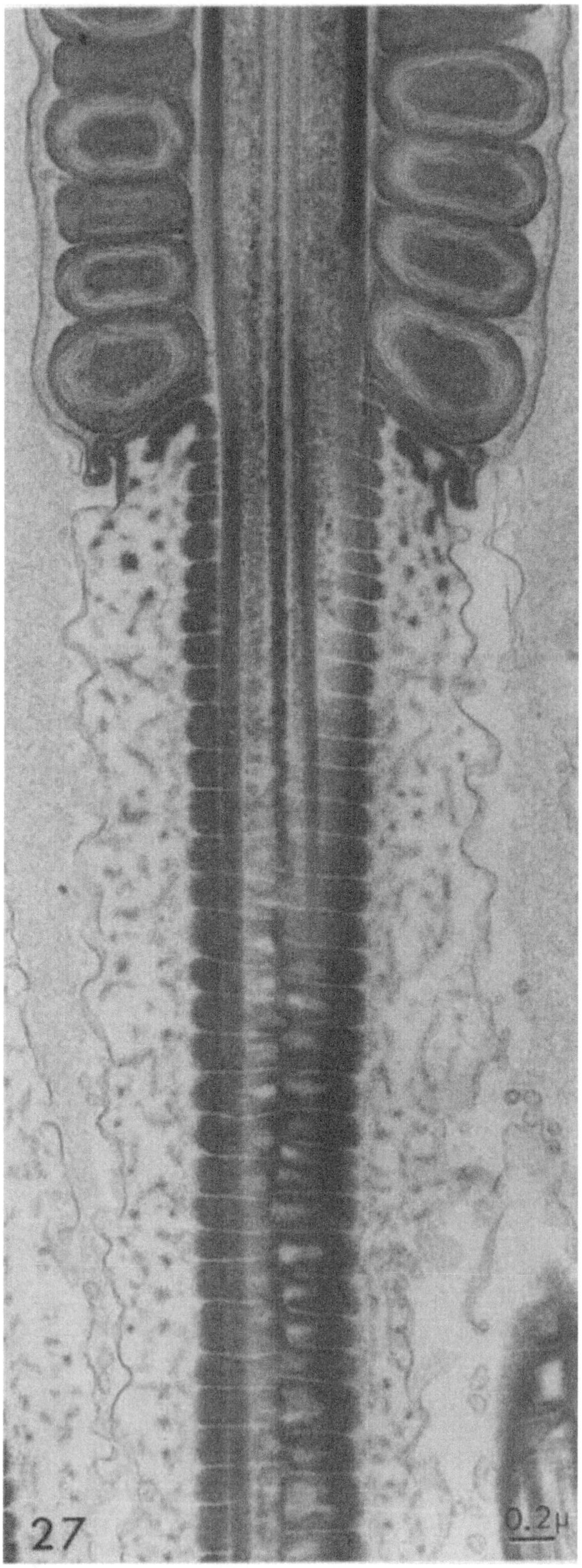


**Fig. 27.** Longitudinal section of a spermatozoon of the North American opossum. The annulus separates the midpiece (top) from the principal piece. In longitudinal section, the fibrous sheath appears segmented. Fuzzy material surrounding the fibrous sheath is characteristic of opossum spermatozoa.

**Fig. 28.** Cross sections through various levels of the principal piece and end piece of late spermatids of the Chinese hamster.

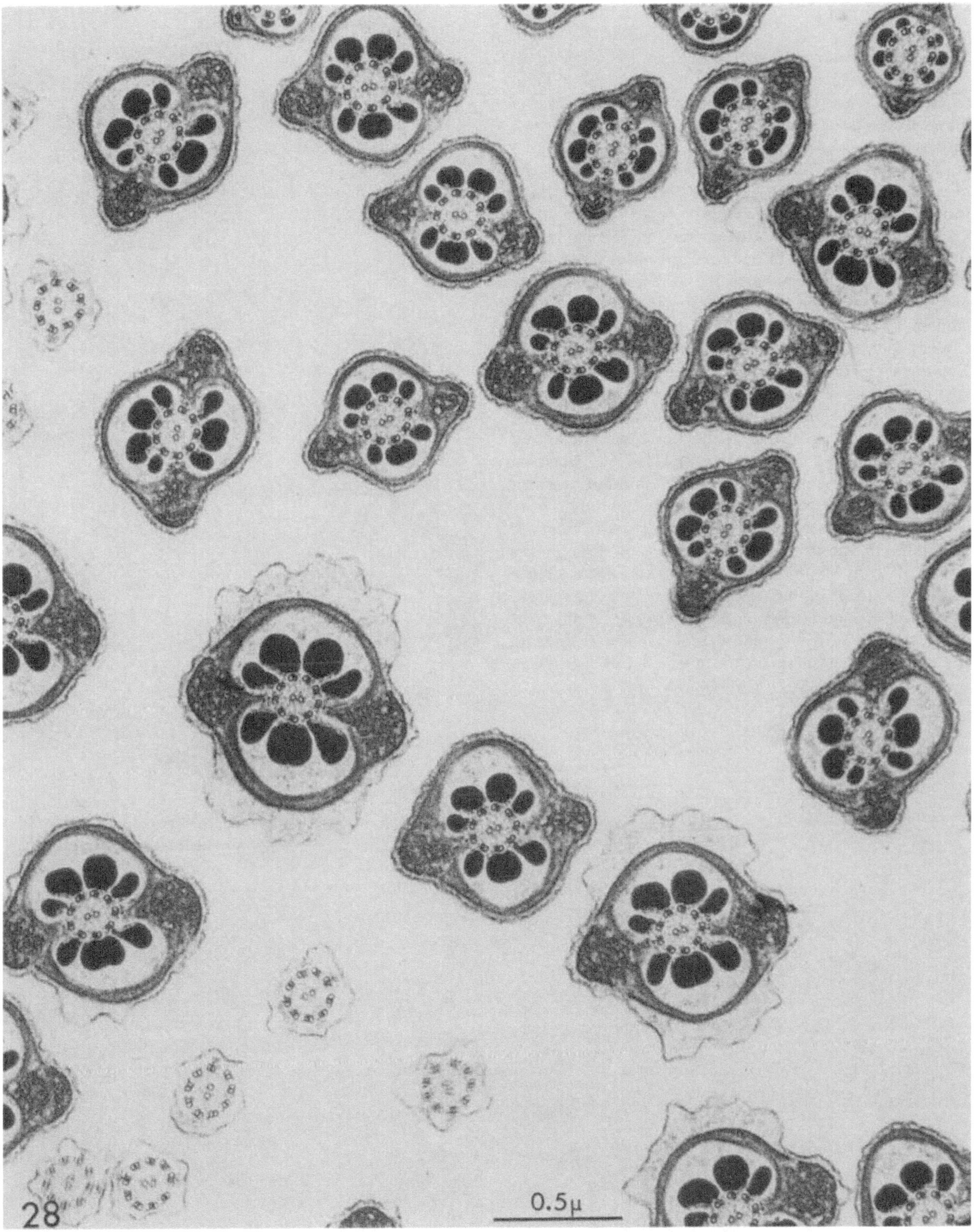
28
0.5μ

## THE INSECT FLAGELLUM

Insect spermatozoa vary greatly in length. Some mayflies produce sperm that are only about 20 $\mu$ long, whereas in some dipterans and hemipterans the sperm may be several millimeters in length. Sperm of most insects have a 9 + 9 + 2 flagellar tubule pattern, that is, the usual 9 + 2 arrangement of flagellar tubules plus nine additional singlet accessory tubules, which form as extensions from subfiber B of each doublet (*31*). The outer nine tubules may be analogous to the dense fibers found in mammalian sperm since dense fibers also form in relation to the nine doublet tubules, are nine in number, and occupy peripheral positions similar to the accessory tubules, (Fig. 29) (*43*).

Generally, two extremely long mitochondrial derivatives occur in mature insect spermatozoa, though some sperm contain only one (Figs 29 and 30). Usually one of the mitochondrial derivatives extends from just behind the nucleus nearly to the posterior end of the cell. The other is somewhat shorter. The mitochondrial derivatives in some species are several hundred microns long and occupy most of the volume of the cell.

The mitochondrial derivative(s) imparts bilateral symmetry to the flagellum. A line drawn through the center of both tubules of the central pair is consistently slightly oblique to the plane of symmetry imparted by the mitochondrial element. Resting guinea pig spermatozoa also have the central pair slightly oblique to the place of flattening of the sperm head (*42*).

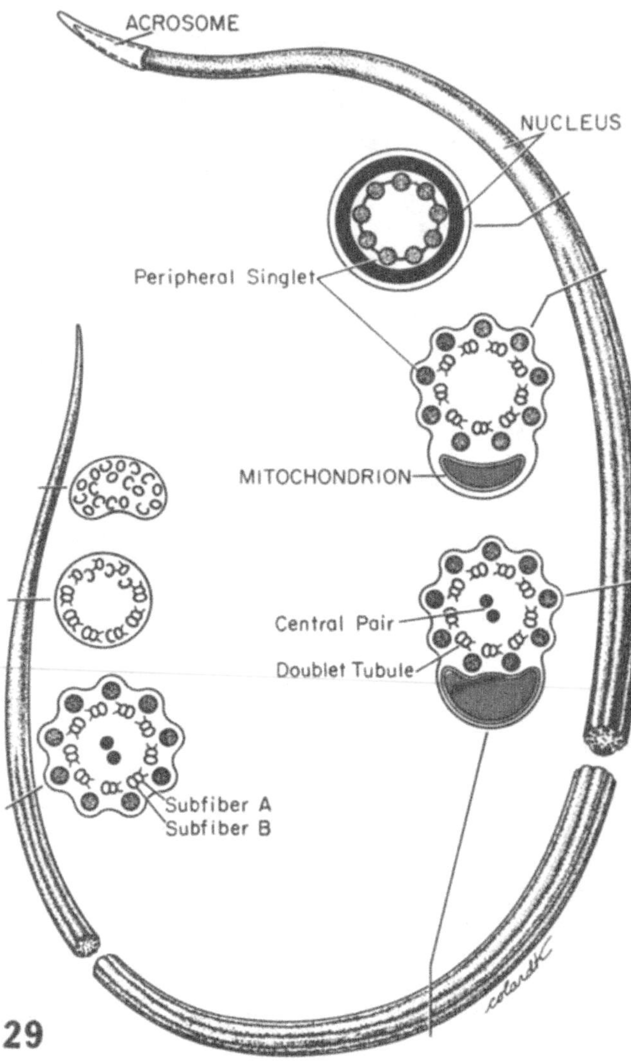


29

Fig. 29. A spermatozoon of the caddisfly *Neuronice* illustrating some of the features common to insect sperm.

Fig. 30. Transverse section, through a cyst of nearly mature spermatozoa of the caddisfly *Macronemum*.

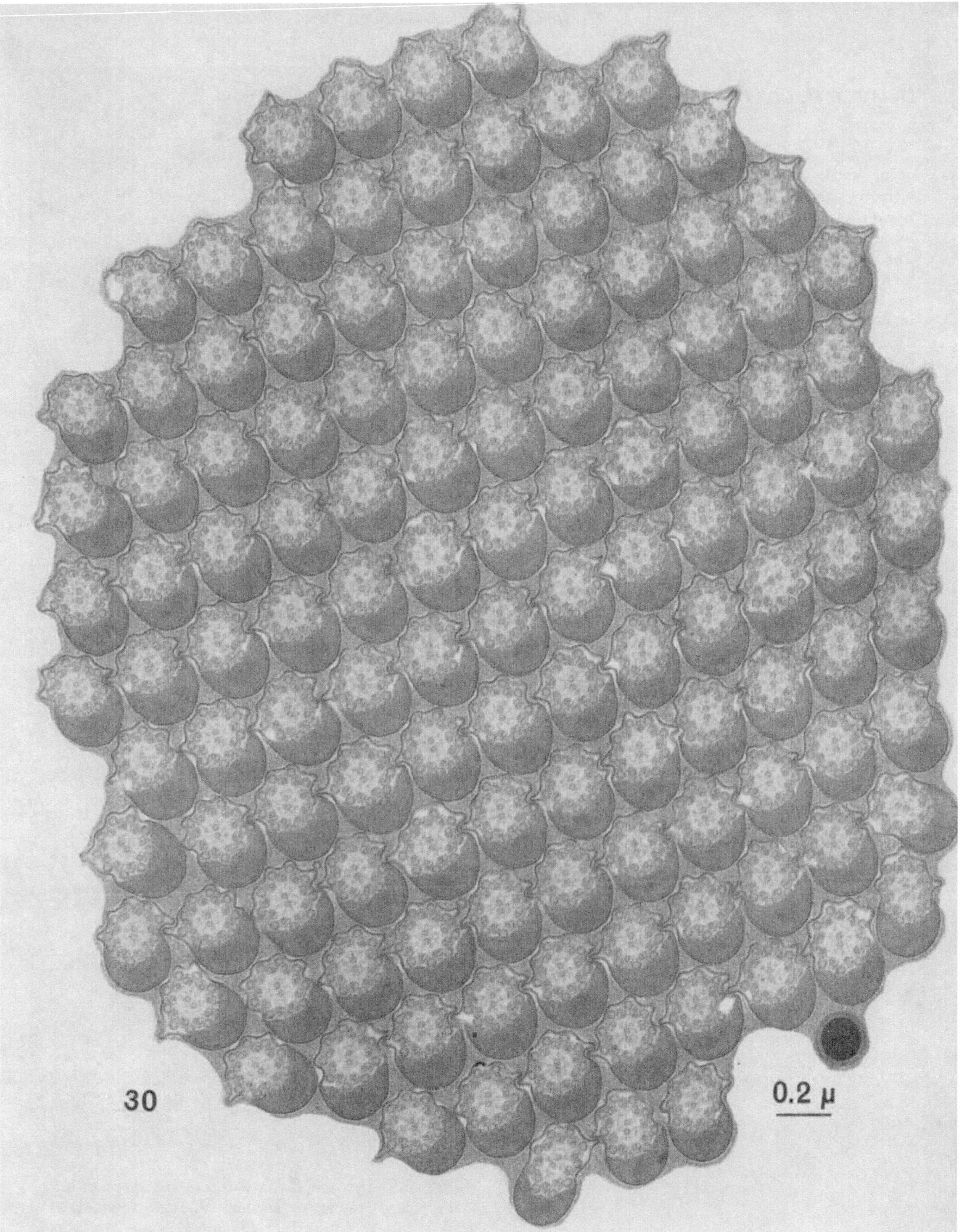


30

## MULTIPLE FLAGELLA

Aberrant multiflagellated spermatozoa commonly occur among normal uniflagellated sperm in the testis of many animals. Retzeus and Meves noted their occurrence some 70 years ago (see *149*) and these spermatozoa have been described by other investigators since. The percentage of multiflagellated spermatozoa is very high in animals with reproductive abnormalities; for example, in sterile mutants of mice (97), X/O *Drosophila* (*77*), or in mutants with defects in cytokinesis (*130*). In transversely sectioned spermatids of the stink bug *Euchistus*, several spermatids have more than one flagellum (Fig. 32). Multiple flagella usually occur in groups of two, four, or eight rather than other numbers. Multiflagellated cells may conceivably arise from one, two, or three abnormal cycles of centriole formation (Fig. 31), or they may arise as a result of incomplete cytokinesis.

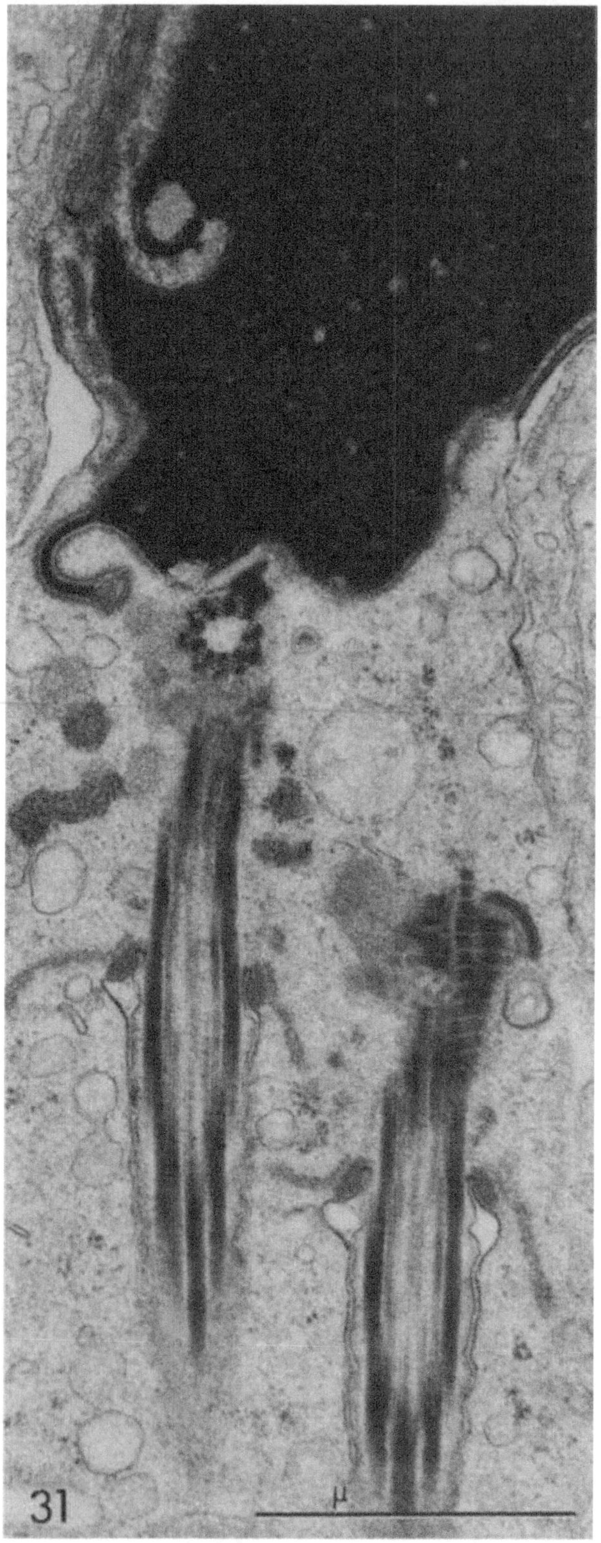


**Fig. 31.** Aberrant biflagellated rabbit spermatid.
**Fig. 32.** Transverse sections through spermatids of the stink bug *Euchistus*. Some of the cells have several flagella.

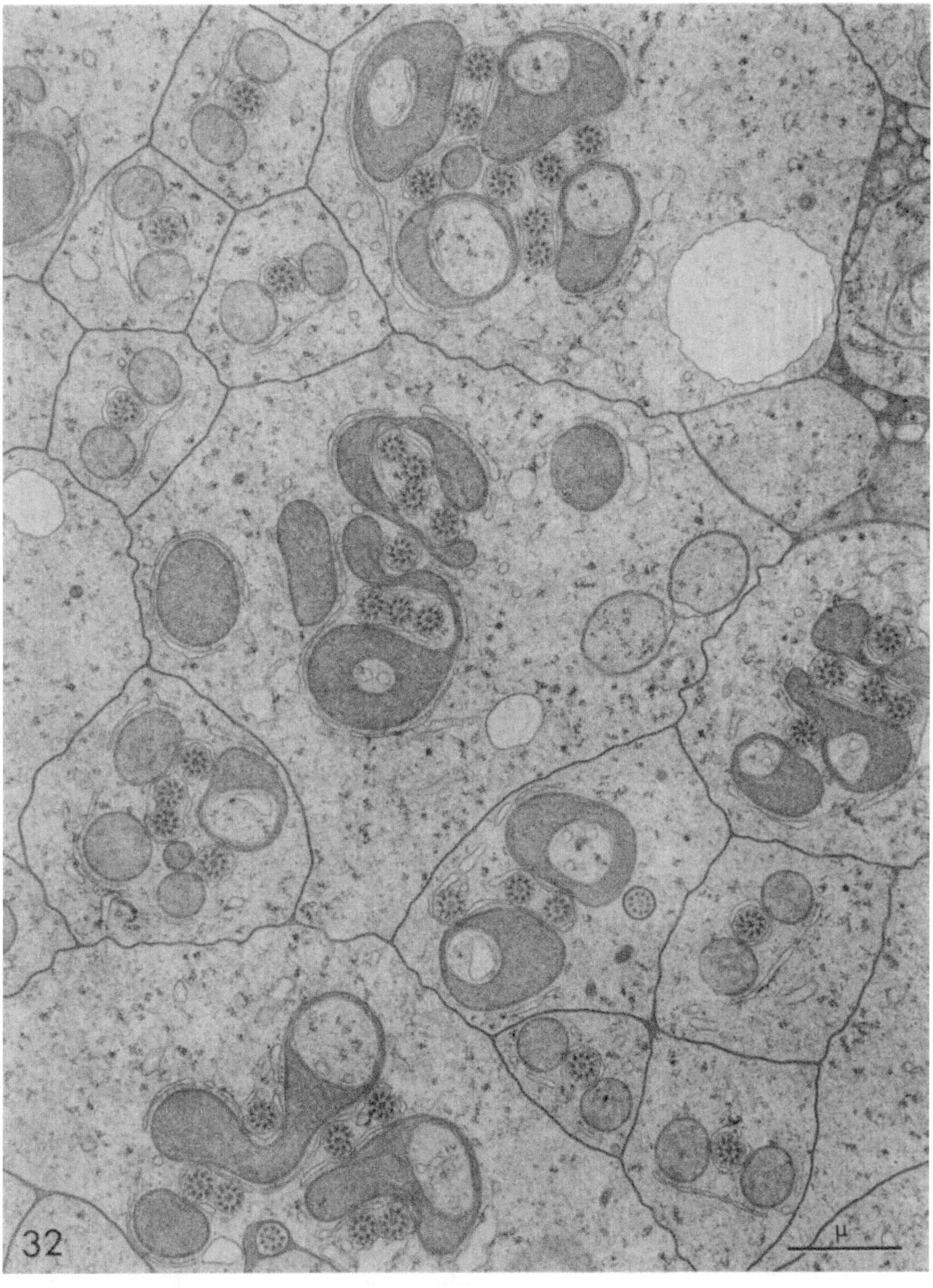
32
μ

## EXCEPTIONAL FLAGELLA

Although spermatozoa of most animals have typical 9 + 2 tubule patterns, there are exceptional forms. Flatworms (*33, 138*), trematodes (*135*), and cestodes (*27*) have a 9 + 1 flagellar tubule array with a single large and elaborate central element. Sperm of some insects possess flagella with 9 + 0, 9 + 1, 9 + 3, and 9 + 7 tubule patterns (Fig. 33) (*101, 103*), and others have flagella in which the nine doublet tubules spiral around the central element or a flagellum tubule that frays out at the end (*101, 104*). In some thrips, the flagellar tubules have no symmetry (*12*). The giant sperm flagella of the Sciaridae contain one of the most striking exceptional tubule patterns. The flagellum of *Sciara coprophila* is composed of 60 to 90 doublet and 60 to 90 associated singlet tubules and develops from a giant centriole also composed of 60 to 90 tubules (*99, 100, 101*). The sperm flagellum of *Rhynchosciara angeli* is even larger (Fig 34) (*136*).

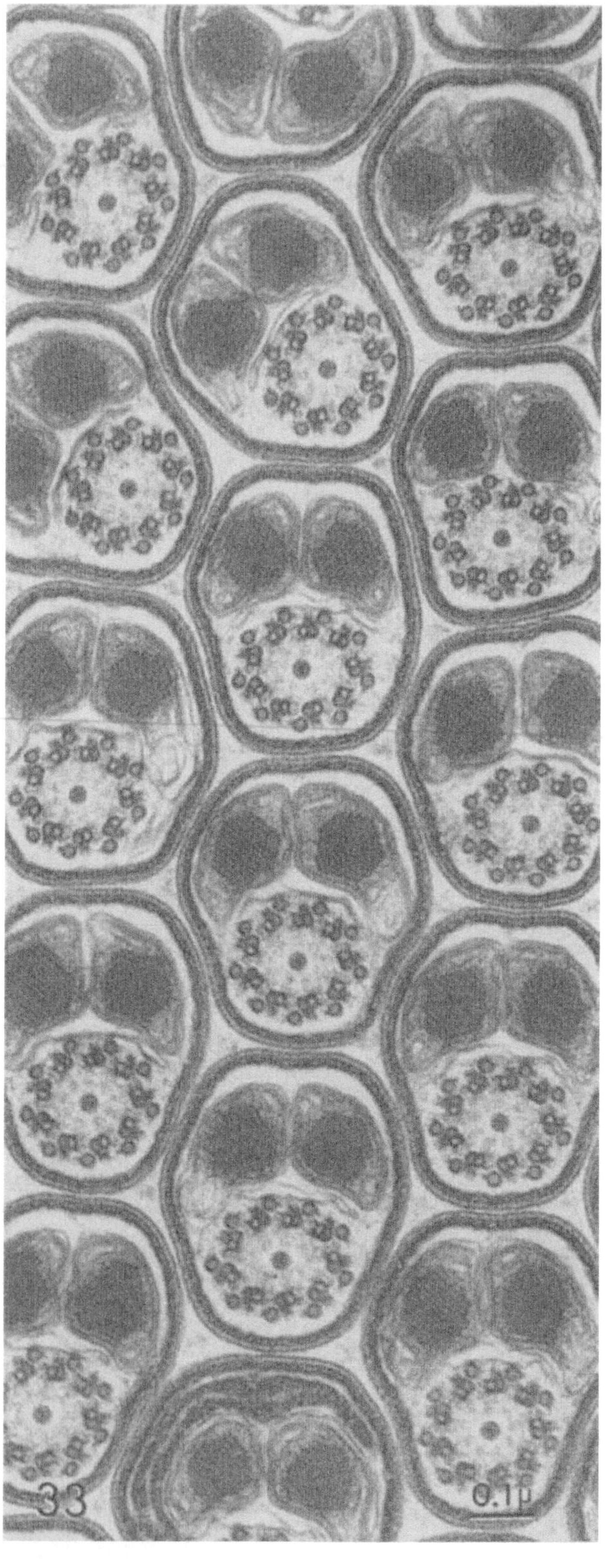


**Fig. 33.** Spermatozoa of the mosquito *Culex* sp. Mosquito sperm have a single central element in the position of the usual central pair.

**Fig. 34.** Transverse section through the sperm flagellum of the fungus gnat *Rhynchosciara.*

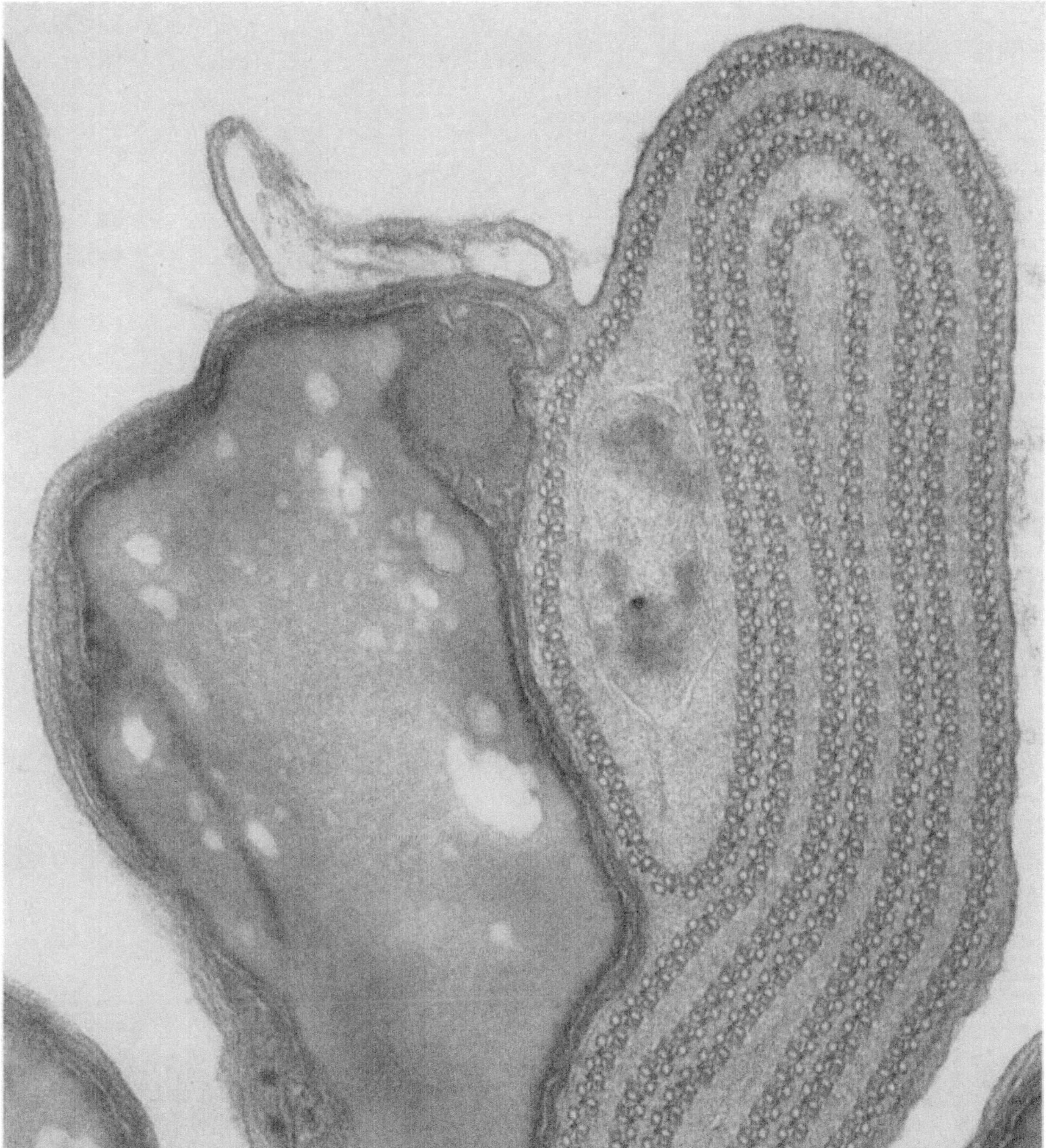
34
0.1μ

## MICROTUBULAR MOTILE APPARATUS

Though they are long thin motile cells, spermatozoa of coccids do not possess a true flagellum (*128, 131*), but have instead a motile apparatus of numerous microtubules. In transverse sections of spermatozoa of the oyster shell scale, *Lepidosaphes ulmi,* microtubules are arranged in a nearly perfect concentric ring around a central core (Fig. 35). At some levels, the sperm nucleus is situated precisely in the center of the microtubular sheath rather than slightly off center. In other coccids, the microtubules are arranged in a spiral (*128*) or in a complicated geometric array (*129*). The development of these interesting gametes has been carefully followed in the light microscope (*68*) and in the electron microscope (*93*). The long spermatozoa form by an outgrowth of papillae into which microtubules and nucleus extend. Papillae are eventually pinched off to form the mature spermatozoa which apparently contain no centriole, acrosome, mitochondria, or mitochondrial derivatives.

**Fig. 35.** Transverse sections of spermatozoa of the oyster shell scale insect *Lepidosaphes.*

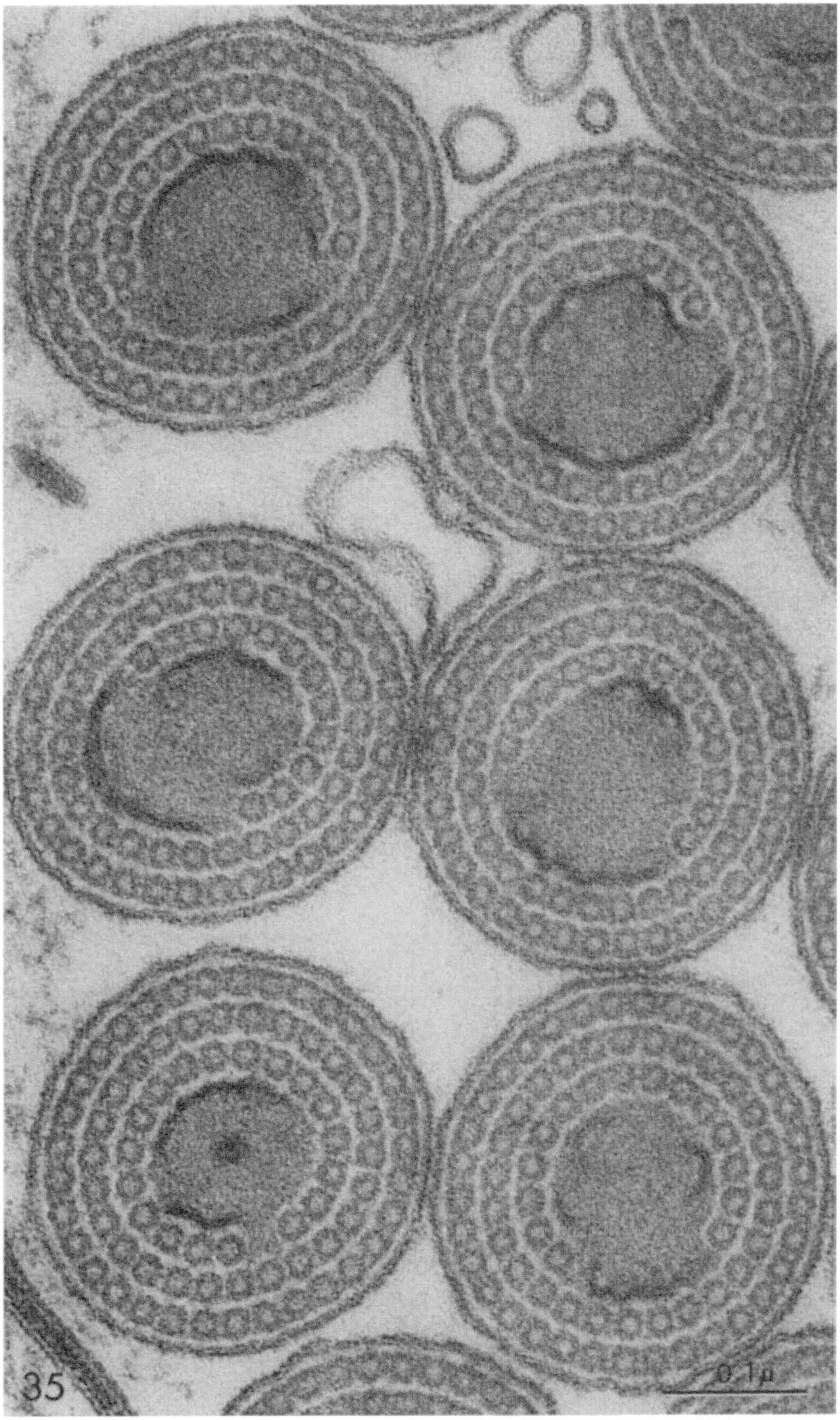
35
0.1 μ

## MITOCHONDRIAL CRISTAE

Mitochondria, like other spermatozoan organelles, assume various shapes during spermiogenesis. In insects, the regularity in the shapes of mitochondria is particularly noticeable (*104*), and the cristae also become regularly disposed (Fig. 37). In some insects, the cristae are not disposed perpendicular to the long axis of the cell but at a regular oblique angle. Sperm mitochondria of most mammalian species do not have well-developed cristae. Mitochondria of opposum sperm are an exception (Fig. 36).

The number of mitochondria in a spermatozoon is characteristic for each species. Marine invertebrates, which shed their spermatozoa into the sea, characteristically have four mitochondria per spermatozoon. Some birds have only one mitochondrion per spermatozoon (*45*) as do certain insects; other insects have two (*104*). Mammalian sperm generally have many more and smaller mitochondria than invertebrates and birds, and the number of mitochondria per spermatozoon varies among mammals. The length of the midpiece and the number of mitochondria within the midpiece appear to be almost constant for all spermatozoa of a given mammalian species (*150, 151*).

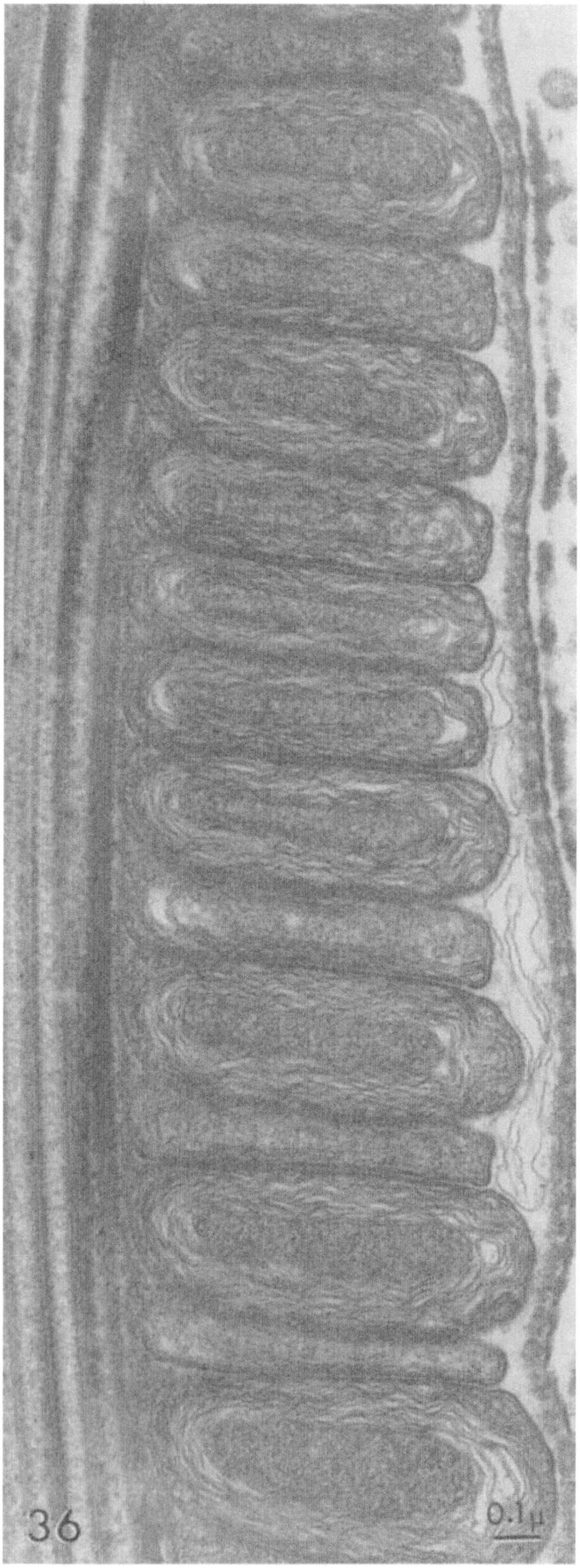


**Fig. 36.** Longitudinal section through a spermatozoon of the North American opossum *Didelphis* showing the numerous intramitochondrial membranes that are elaborated during spermiogenesis in opossums.

**Fig. 37.** Longitudinal sections through spermatozoa of a checkered beetle *Cleridae* showing regularly arranged cristae.

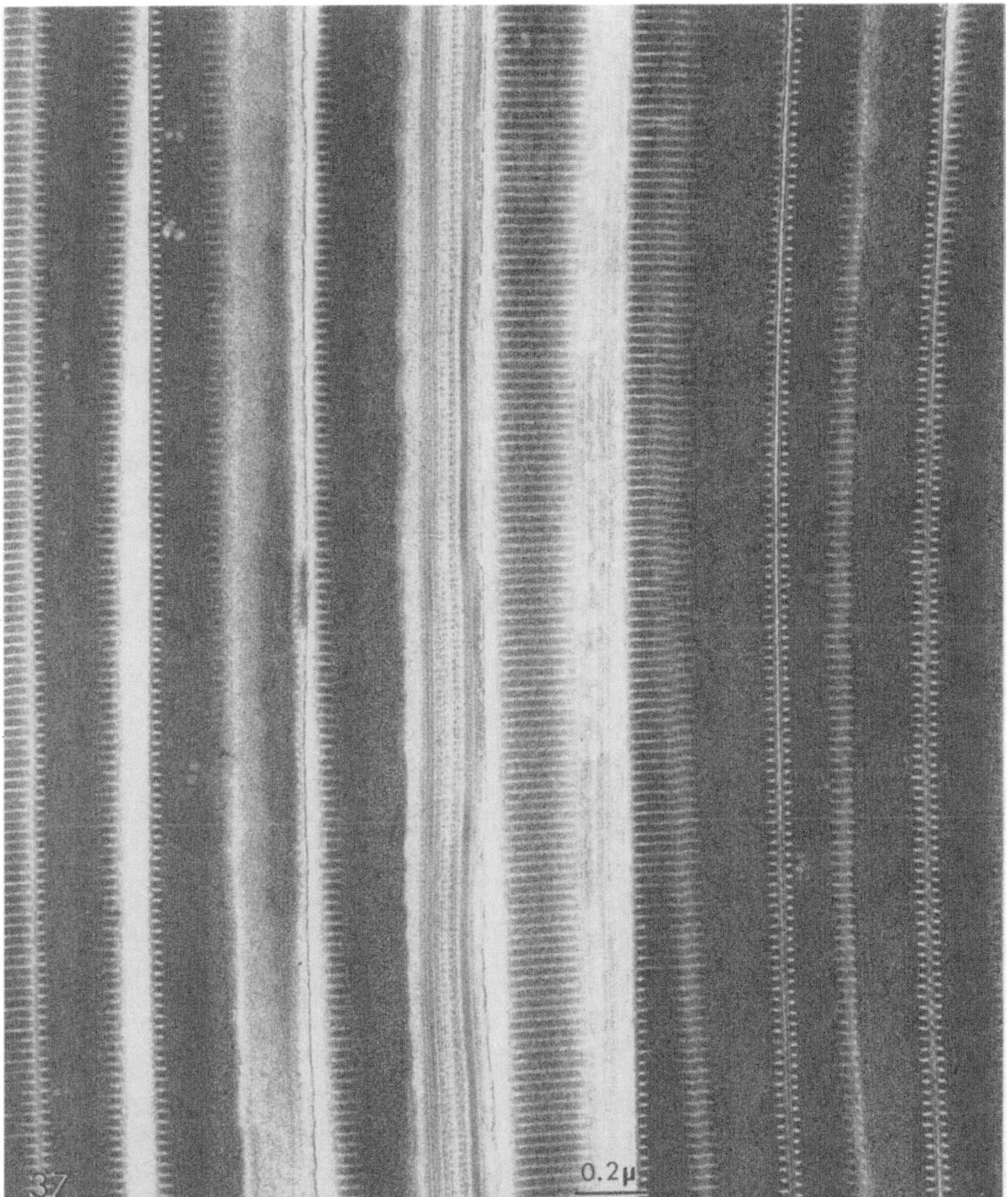
37
0.2μ

## MITOCHONDRIAL DISTRIBUTION

The number, shape, and location of mitochondria in mature sperm is generally very nearly identical in mature spermatozoa of a given species. The ordering of the array of mitochondria apparently begins early in spermiogenesis. In young mammalian spermatids, numerous mitochondria are often nonrandomly distributed within the cytoplasm. In some mammals, mitochondria all cluster in one small region of the spermatid (Fig. 38). In young spermatids of other species (Fig. 39), mitochondria occur at the periphery of the cell just below the plasma membrane. In both examples (Figs. 38 and 39), the lumen of the testis tubule is toward the top; more mature spermatids occur closest to the lumen, and spermatogonia are seen at the bottom of the picture (near the basal side of the tubule).

The distribution of mitochondria is drastically altered in later stages of spermiogenesis when the annulus moves from its position just posterior to the basal body to a position that defines the end of the midpiece. At that time, mitochondria take up positions in the midpiece along the spermatid flagellum (Fig. 43).

**Fig. 38.** Photomicrograph of a portion of a rabbit seminiferous tubule. Toluidine blue stained 1 $\mu$ thick Epon section.

**Fig. 39.** Photomicrograph of a portion of a kangaroo rat seminiferous tubule. Toluidine blue stained 1 $\mu$ thick Epon section.

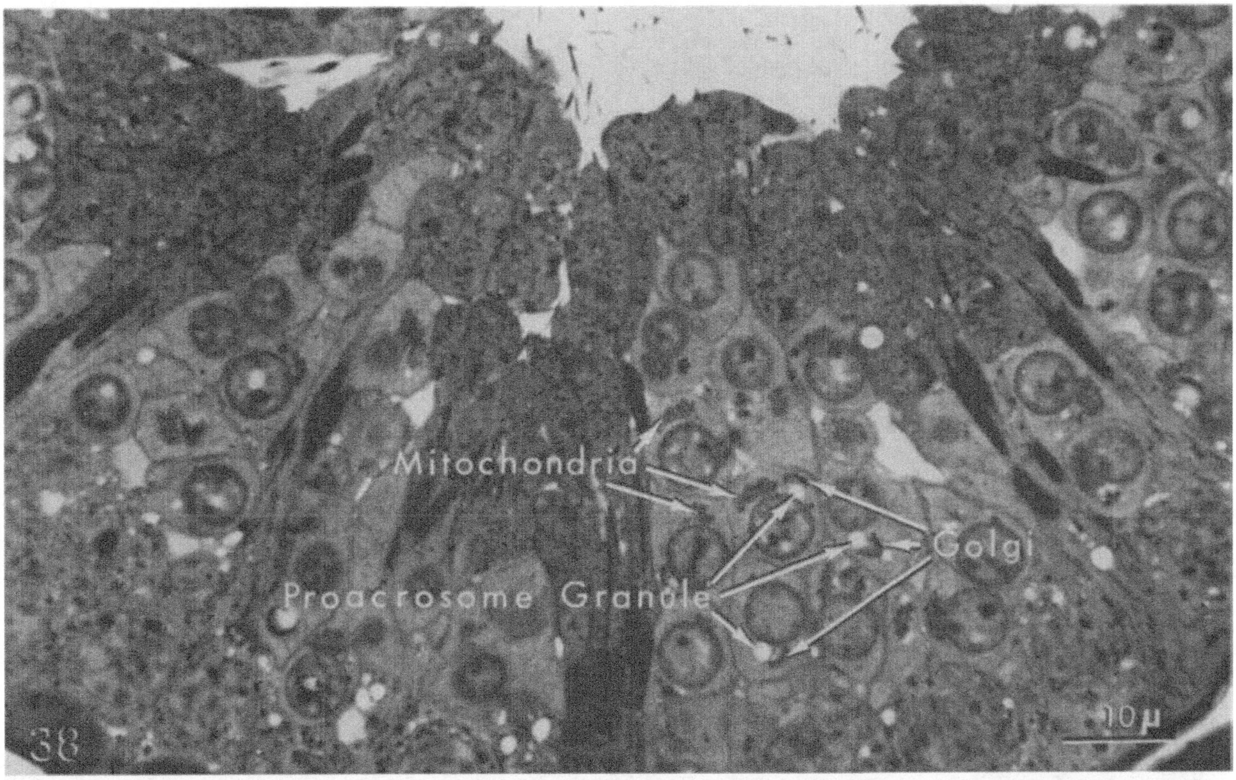
Mitochondria
Golgi
Proacrosome Granule
10 μ
38

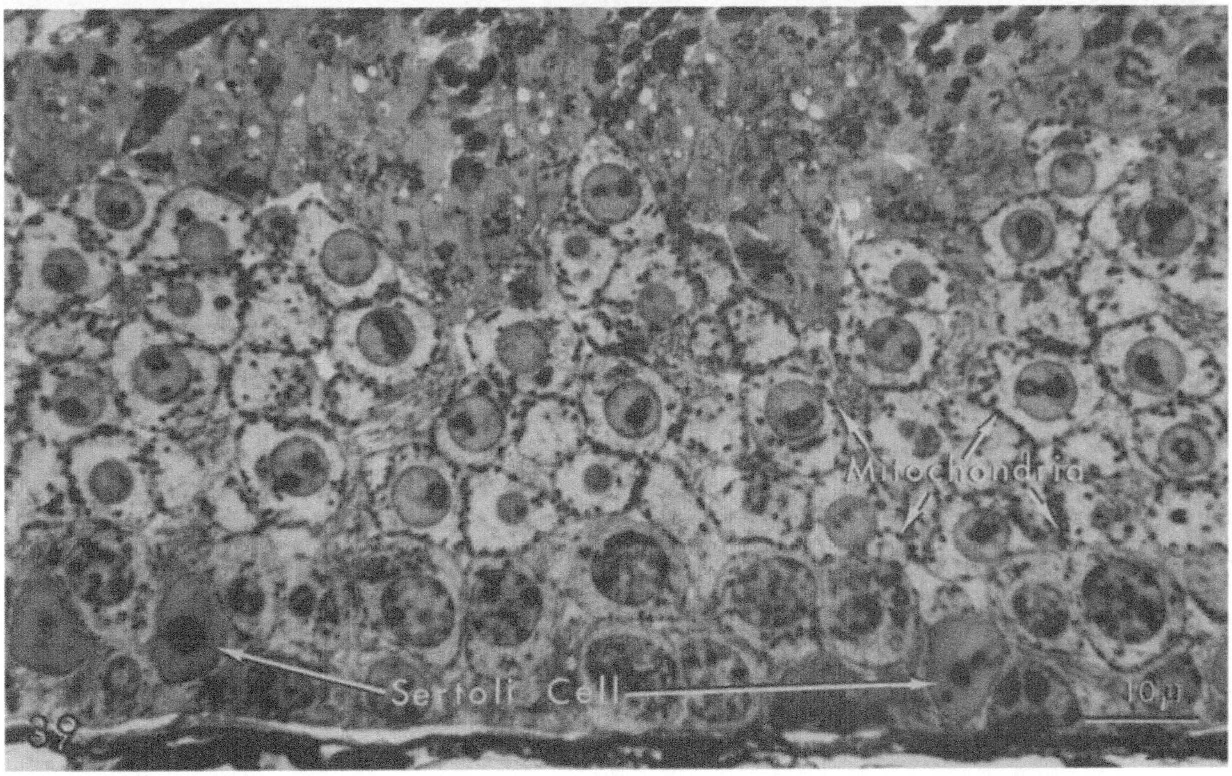
Mitochondria
Sertoli Cell
10 μ
39

## MITOCHONDRIAL INCLUSIONS

Spermatozoan mitochondria contain many types of inclusions. Glycogen is commonly found in mitochondria of many invertebrates (*2, 8*). [Extramitochondrial glycogen also occurs in spermatozoa of some species (*58, 83, 140*).] Paracrystalline inclusions, presumably proteinaceous, accumulate during spermiogenesis in the mitochondria of many insects (*91, 104, 146*). In some species, intramitochondrial materials accumulate to such an extent that they eventually fill most of the volume of the cell. Some mammalian spermatozoa contain material in the intercristae space (Fig. 40). Whether such mitochondrial inclusions are enzymatic or storage material or serve some other function is not known.

**Fig. 40.** Longitudinal section through the midpiece of a kangaroo rat spermatozoon.

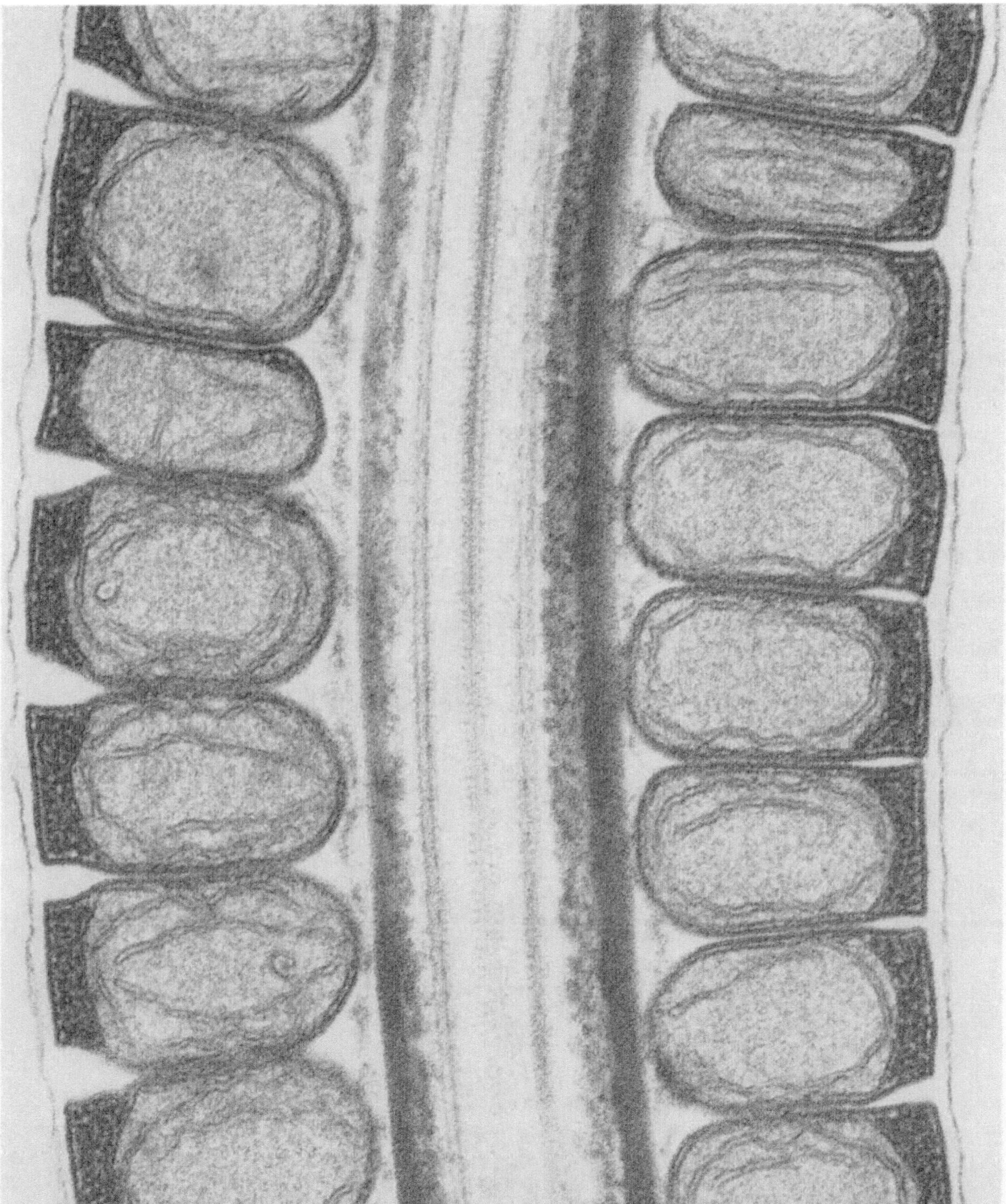


40

## THE NEBENKERN

Very early in spermiogenesis in insects (*104, 113*), some snails (*11, 71*), and ostracods (*121*), all the mitochondria aggregate in one region of the cell and fuse into one mass. This large ball of mitochondria is called the nebenkern. Using serial section analysis, Pratt (*113*) followed the fusion of mitochondria in spermatids of the stink bug *Murgantia*. She found that the mitochondria within the nebenkern undergo complicated rearrangements and configurational changes. Eventually the nebenkern unravels and separates into two large mitochondrial derivatives. These two huge mitochondria then elongate along the length of the flagellum, following the posterior migration of the ring centriole, a structure that is analogous to the annulus of mammalian sperm (*104*). In Fig. 41 of a young spermatid of *Euchistus* (stink bug), the nebenkern is located in its characteristic position near the base of the growing flagellum. The bulbous appearance of the tip of the flagellum, which is only a few microns long at this stage, is typical of growing flagella. Material near the basal body is the forming centriole adjunct which eventually extends around the basal body and may serve to anchor the flagellum onto the nucleus. The centriole adjunct in insect sperm thus may have a function analogous to that of the columns of the connecting piece in mammalian spermatozoa. The nuclear membrane is regular and thickened in the region of the flagellar base, an appearance characteristic of spermatids of virtually all animals. There is a small ring centriole at the base of the flagellum.

**Fig. 41.** Young spermatid of the stink bug *Euchistus*.

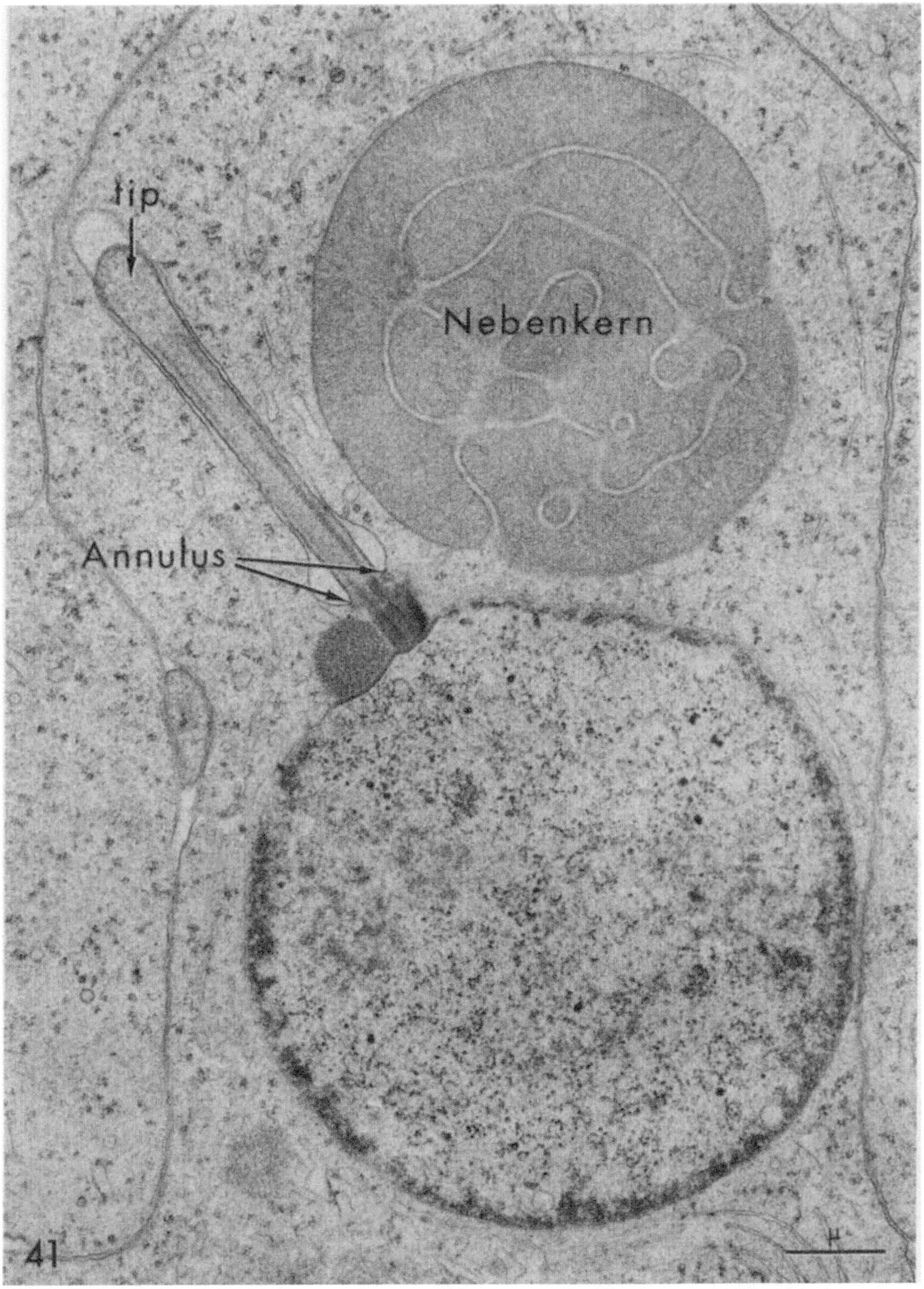
tip
Nebenkern
Annulus
41
μ

## THE ANNULUS

The annulus of mammalian spermatids is originally situated just posterior to the basal body where it seems to anchor the involuted plasma membrane. The Chinese hamster spermatid in Fig. 42 has been transected through the annulus, revealing a bipartite structure. This annulus consists of a central mass of circularly arranged fibers and an outer network of flocculent material. This appearance characterizes the annulus of young spermatids; the outer material disappears during spermiogenesis. In longitudinal sections, the annulus appears as two dense dots on either side of the flagellum with the less conspicuous flocculent material trailing peripherally (see Fig. 44). Later the annulus moves down the axial filament complex until it comes to the fibrous sheath where it stops, forming a boundary between the midpiece and principal piece.

The annulus is believed to derive in part from a large fibrous electron-dense structure, termed the chromatoid body (*44*). The origin and constitution of the chromatoid body itself is controversial. It has been suggested that the material that composes the chromatoid body is of nuclear origin and passes through nuclear pores (*37, 85*), or that the chromatoid body arises from cytoplasmic-dense material that aggregates in association with mitochondria (*46*). Some evidence supports the idea that the chromatoid body contains RNA (*85*) while other evidence suggests that it does not (*41*).

**Fig. 42.** Transverse section through young Chinese hamster spermatids. The cell in the center has been cut through the annulus. Cells at the lower left and right are transected at the level of the principal piece.

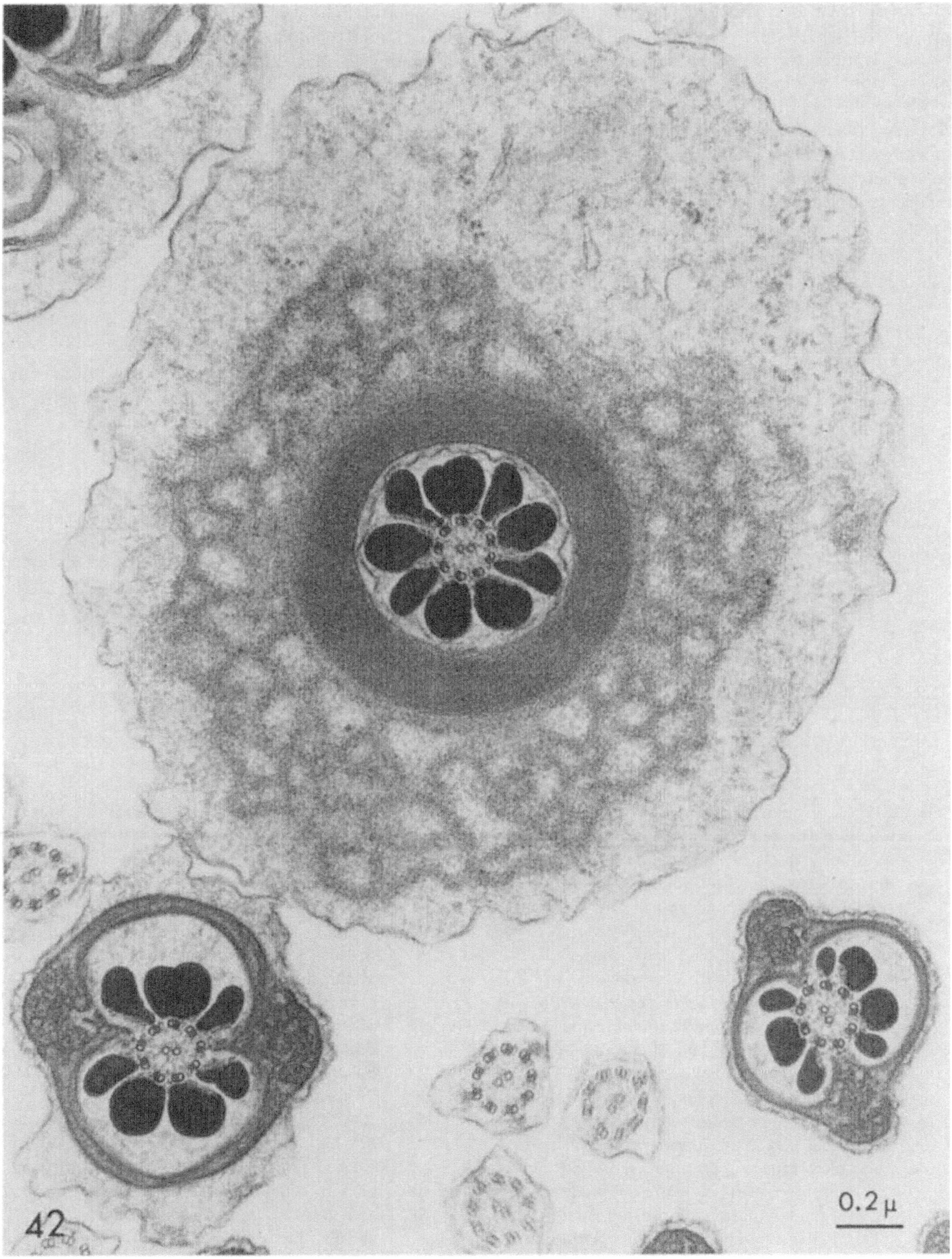
42
0.2 μ

## MOVEMENT OF THE ANNULUS

Until fairly late in spermiogenesis, the annulus of mammalian spermatids is situated just posterior to the columns of the connecting (Fig. 44). The involuted plasma membrane is connected to the annulus. Figure 45 shows a later stage in which the annulus has moved posteriorly to the level of the fibrous sheath. This process is shown diagrammatically in Fig. 43. During the posterior migration of the annulus, the mitochondria take up positions adjacent to the flagellum, and the outer flocculent zone of the annulus disappears. The position of the fibrous sheath may determine the length of the midpiece by physically stopping the posterior movement of the annulus. In younger spermatids (Fig. 44), a tubular complex (*115*) occurs just anterior to the forming fibrous sheath. The tubular complex is a transient structure not seen in mature spermatozoa. In Figs. 44 and 45, the nuclear membrane on the posterior edge of the nucleus loops into the cytoplasm (arrows). The outfolded nuclear membrane contains no chromatin. Nuclear membrane is believed to be sloughed off in this manner (*116*).

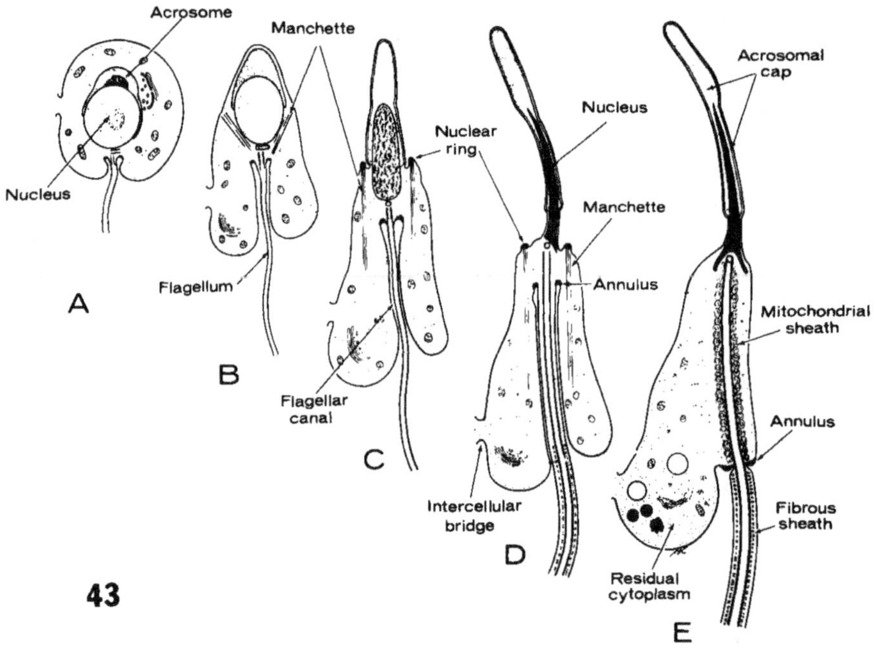


**43**

**Fig. 43.** Stages of spermiogenesis in the guinea pig showing the position of the annulus as well as other aspects of mammalian spermiogenesis discussed in this monograph. (A) The nucleus and acrosomal cap of the spermatid are centrally located in the spermatid cytoplasm in the Golgi and cap phases of development. (B) With completion of the acrosomal cap, the Golgi migrates into the caudal cytoplasm, microtubules appear tangential to the caudal half of the nucleus, and spermatid elongation begins. (C) Caudal displacement of cytoplasm brings the plasma membrane into close apposition to the acrosomal cap. Nuclear condensation begins; the microtubules become arranged in a cylinder originating in a dense nuclear ring at the caudal margin of the acrosomal cap. (D) Nuclear condensation and spermatid elongation continue. Nuclear ring moves back to the level of the base of the flagellum; flagellar canal lengthens and fibrous sheath begins to form around axoneme of the free portion of the flagellum. (E) Nuclear ring and manchette disappear; annulus migrates caudally to the anterior margin of the fibrous sheath; mitochondria gather around the base of the flagellum anterior to the annulus and form the mitochondrial helix of the middle piece. Diagram and legend from *Develop. Biol.* **26**, 220 (1971). Courtesy of Drs. D. W. Fawcett, W. A. Anderson, and Dr. D. M. Phillips.

**Fig. 44.** Longitudinal section of a rabbit spermatid. At this stage of development, the annulus is located just posterior to the neck region.

**Fig. 45.** Longitudinal section of a rabbit spermatid at a later stage than shown in Fig. 44. The annulus has migrated posteriorly to the forming fibrous sheath.

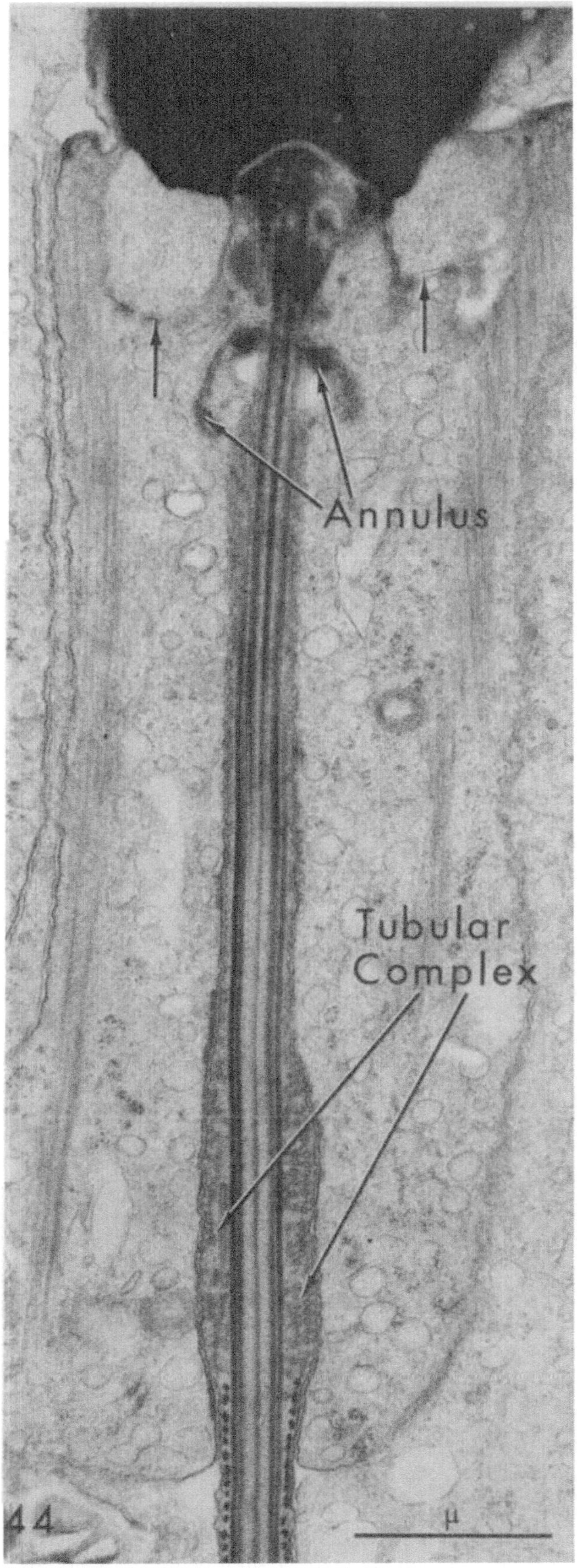
Annulus
Tubular
Complex
44
μ

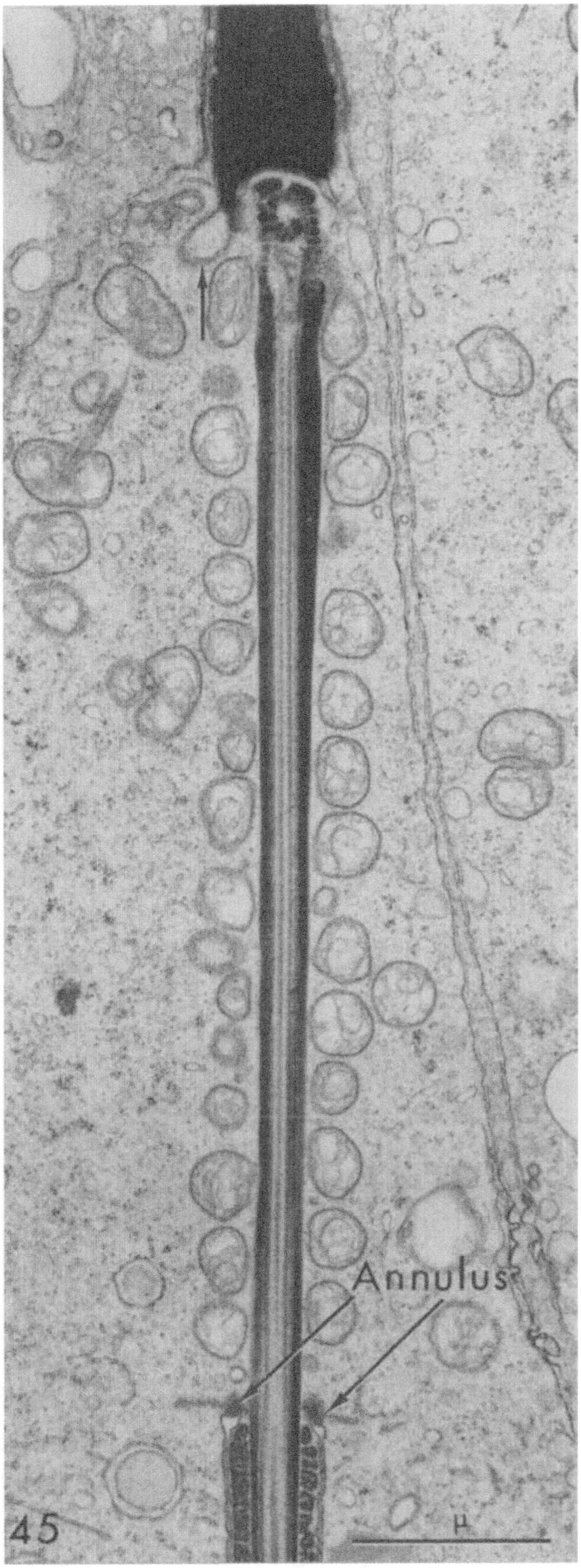
Annulus
45
μ

## CYTOPLASMIC SLOUGHING

In spermatozoa of most animal species, the volume of spermatids is reduced severalfold during spermiogenesis. Ribosomes, ER, vesicles, Golgi complexes, and microtubules are lost. This process occurs in the absence of a lysosomal system as is generally associated with cytoplasmic degradation in other cell types. Instead, large areas of spermatic cytoplasm containing unwanted structures are eliminated by being pinched off from the spermatid (Fig. 46). These residual bodies are strongly basophilic due to the high concentration of ribosomes within them. In mammals, Sertoli cells apparently aid in the sloughing of cytoplasm surrounding and attaching to the residual body; the cytoplasmic connection between residual body and spermatid is severed when the nearly mature spermatozoon is released into the lumen of the seminiferous tubule (*50*). Lysosomes, present in the residual bodies, apparently function in degradation of the residual body after it becomes dissociated from the spermatozoon (*39*). In insects, large spheres of cytoplasm containing superfluous material are pinched off from the spermatid and appear free in the cyst lumen (Fig. 47).

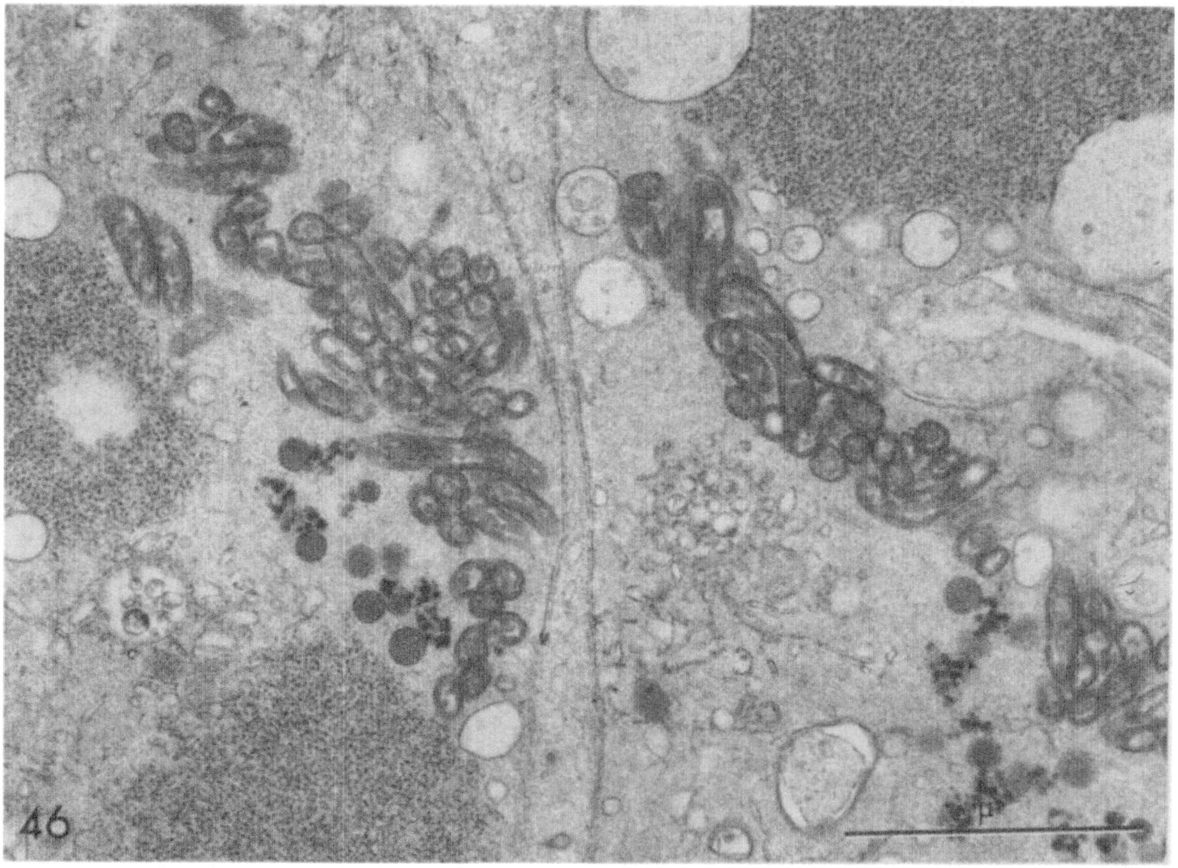


**Fig. 46.** Portions of two residual bodies of rabbit spermatids. Excess mitochondria and large masses of ribosomes characterize residual bodies in mammals.

**Fig. 47.** Longitudinal section of spermatids of the dobson fly *Corydalis*. Large regions of cytoplasm containing rough endoplasmic reticulum are apparently being sloughed into the cyst lumen.

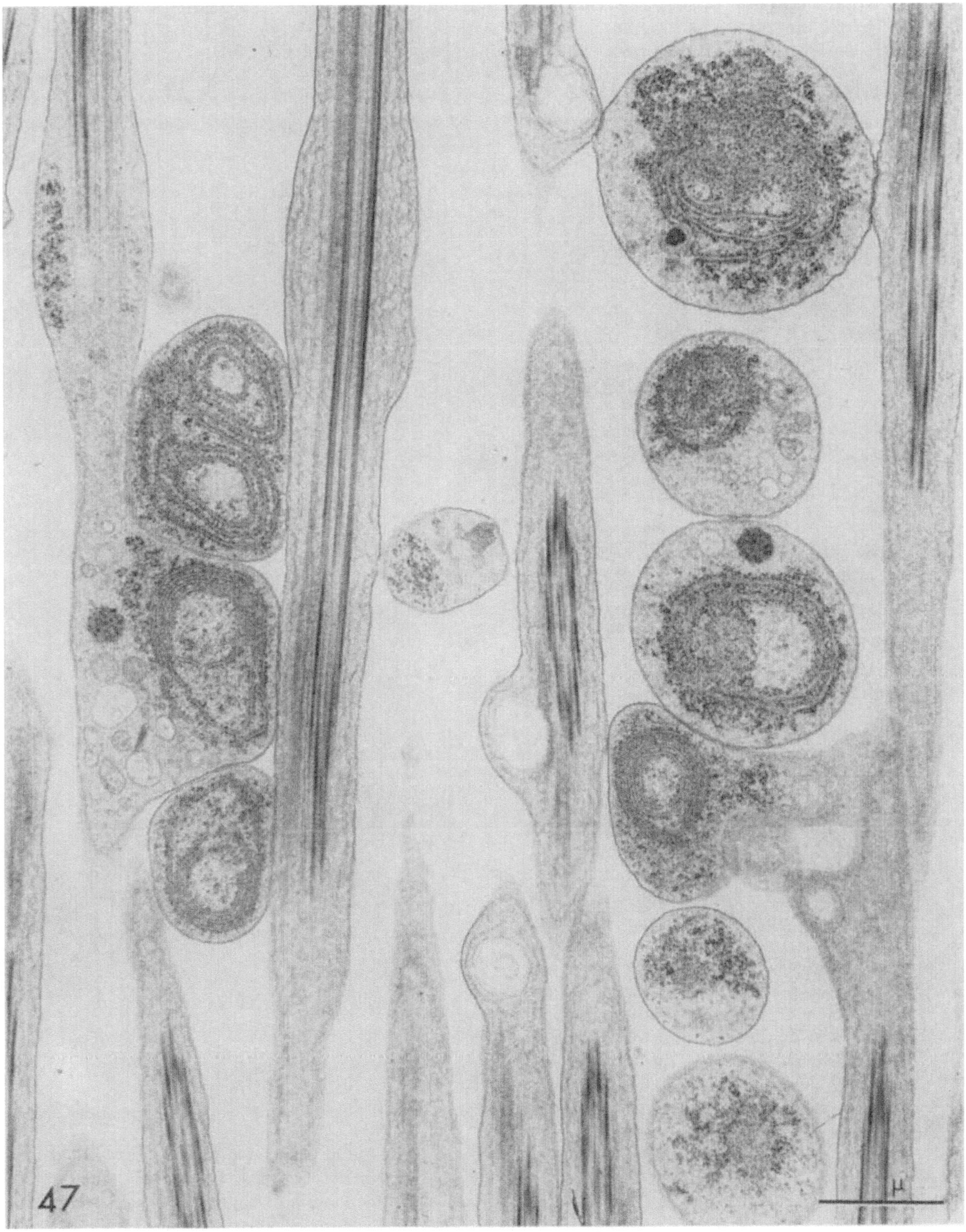
47
μ

## THE PROTOPLASMIC DROPLET

One of the interesting aspects of sperm development is that it is often not complete when the spermatozoa leave the testis. For example, spermatozoa of some insects undergo pronounced ultrastructural alteration in the female genital tract (*101, 104*), and mammalian spermatozoa may undergo changes in the shape of the head (*47, 50*) or, in the case of opossums, may form pairs during passage through the epididymis (Fig. 48) (*105*). The protoplasmic droplet is a mass of cytoplasm surrounding the neck region of mammalian spermatozoa when they leave the testis. The droplet is characterized by flattened vesicles (Fig. 49). As the spermatozoa pass through the epididymis, the droplet moves down the midpiece to a position just anterior to the annulus. The droplet subsequently detaches from the spermatozoon.

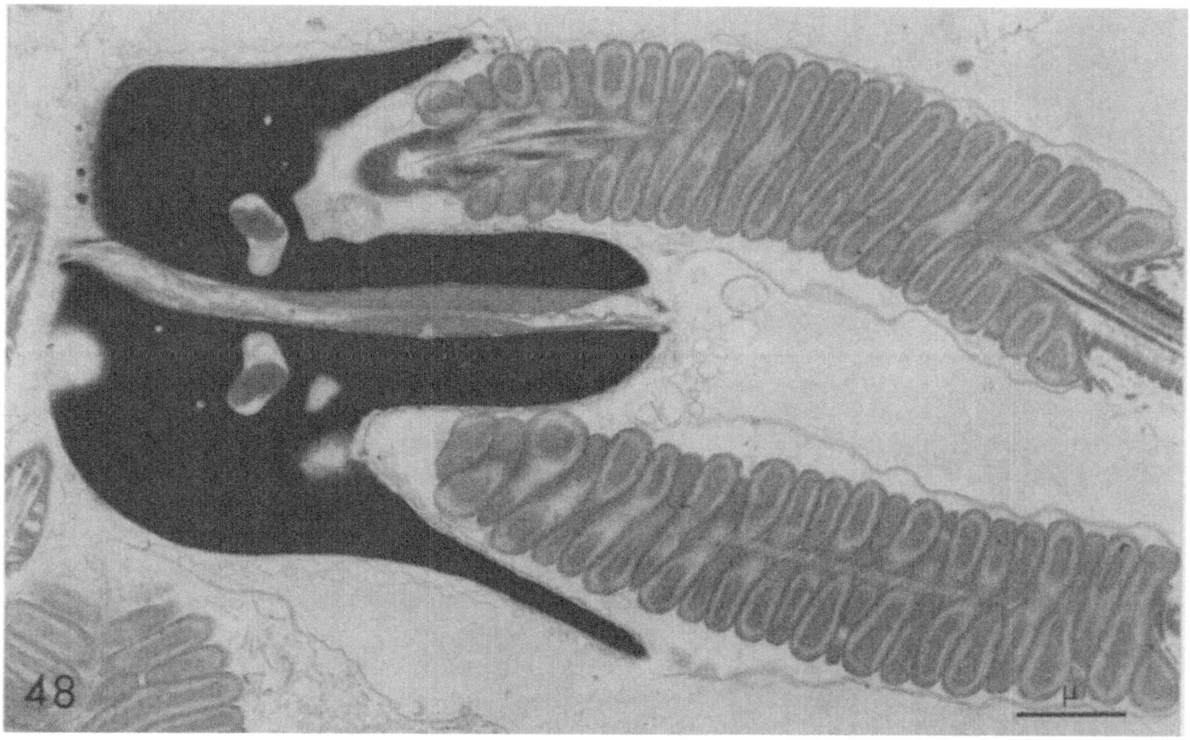


**Fig. 48.** Longitudinal section of two paired spermatozoa of the North American opossum *Didelphis.* Sperm conjugation occurs in American marsupials and in certain insect species (*149*). In American marsupials, spermatozoa are released from the testis singly and pair in the epididymis. Flattened acrosomes are situated between two electron-dense clothespin-shaped nuclei. Paired opossum sperm swim in a coordinated pattern (*107*). The function of pairing is not understood, but the position of the acrosomes between conjugated nuclei suggests that pairing may in some way serve to protect acrosomes.

**Fig. 49.** Longitudinal section of two Chinese hamster spermatids showing the protoplasmic droplet. Mitochondria of Chinese hamster spermatids are filled with dense material (arrows) which disappears during spermiogenesis.

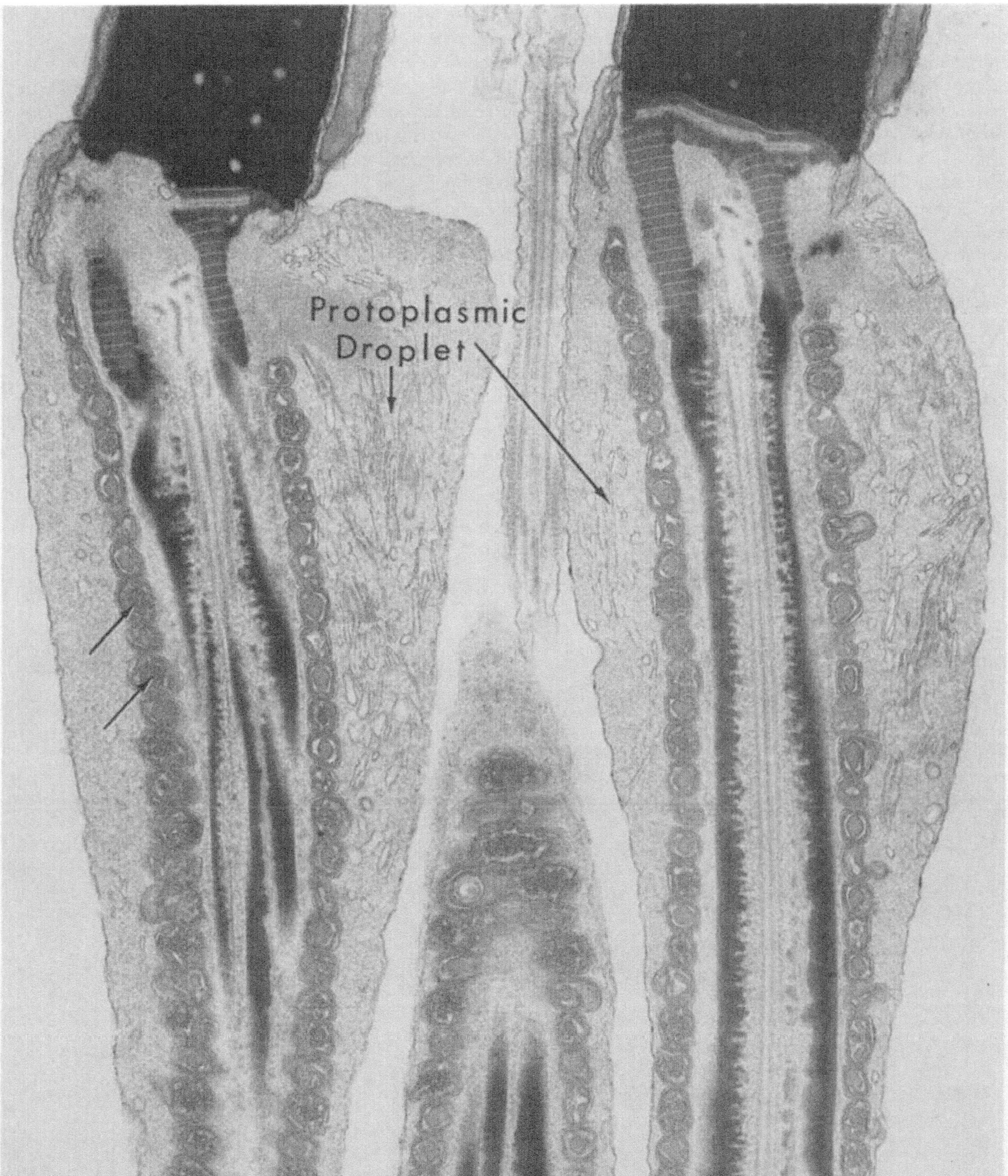
Protoplasmic
Droplet
49
μ

## THE SPERM SURFACE

Many investigators believe that changes in properties of the spermatozoan cell surface may be involved in the physiological change that sperm of some mammals undergo in the female genital tract. This change, termed capacitation, renders spermatozoa capable of fertilization. The change to a capacitated state has not been correlated with any morphological change which can be demonstrated with standard EM procedures. Hopefully, future cytochemical studies will be helpful in elucidating the role of the cell surface in capacitation. Small vesicles and tubules overlie the plasma membrane of sperm heads of some mammalian species (Fig. 14). Their position seems to be correlated with a "sticky" portion of the sperm head (*108*), and vesicles may possibly be involved in sperm egg interactions. Special stains have been used to demonstrate orderly structures (2) and the distribution of negative charges (*151a*) on sperm surfaces.

Lepidopteran sperm are covered with material termed lacinate appendages (*10, 106, 126*). These beautiful bladelike structures occur on the outside of the plasma membrane. The number, size, and shape of these appendages vary among species. One of the appendages, termed the reticular appendage, is consistently morphologically dissimilar from the others (Fig. 51). The position of this appendage with respect to internal characteristics of the flagellum is constant. The other appendages vary in number and shape among different Lepidopteran species. In transverse section, they appear to be composed of alternating electron-lucid and electron-dense areas with a 90 Å periodicity (Fig. 52). The spermatids are embedded in one of the large polytene cells which compose the cyst wall. The lacinate appendages do not persist but are replaced or modified into another morphological type of extracellular material as spermatozoa leave the testis, and they are further modified in the female genital tract (*106*). Other types of extracellular material occur on the plasma membrane of other insects (Fig. 50) (*13, 74*).

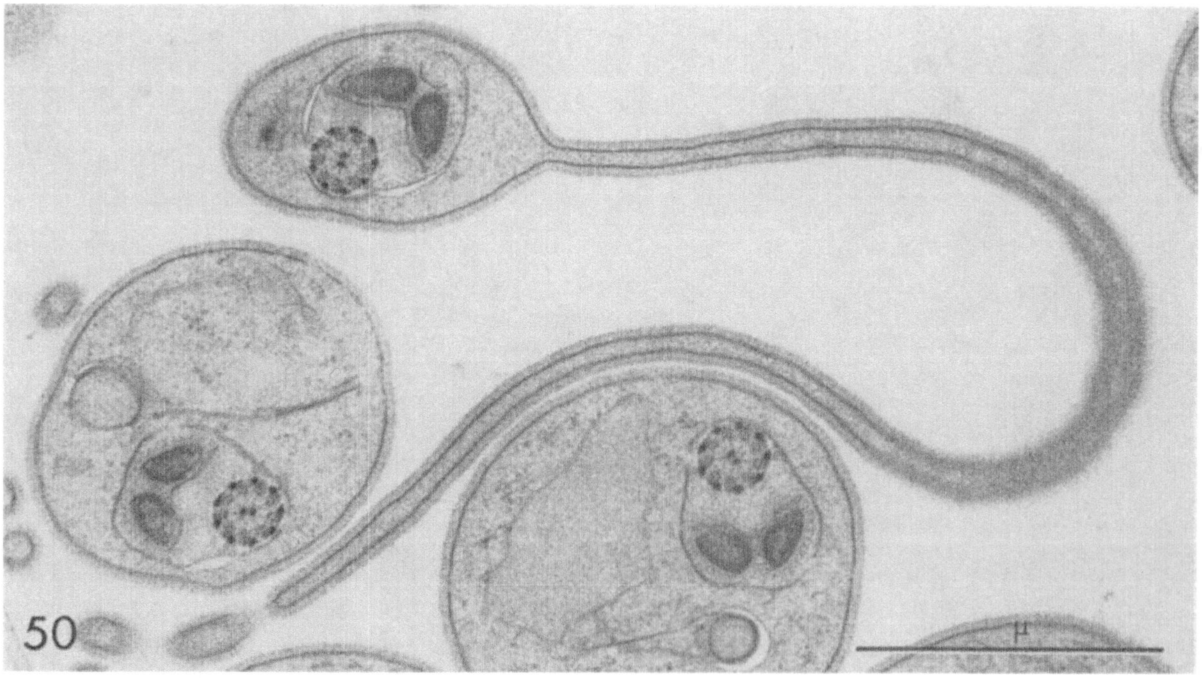


**Fig. 50.** Transverse section of spermatids of the spur-throated grasshopper *Melanoplus*. Material coats the plasma membrane. Excess cytoplasm is apparently being sloughed from one spermatid.

**Fig. 51.** Transverse section through nuclei of spermatid of the geometrid moth *Haematopis*.

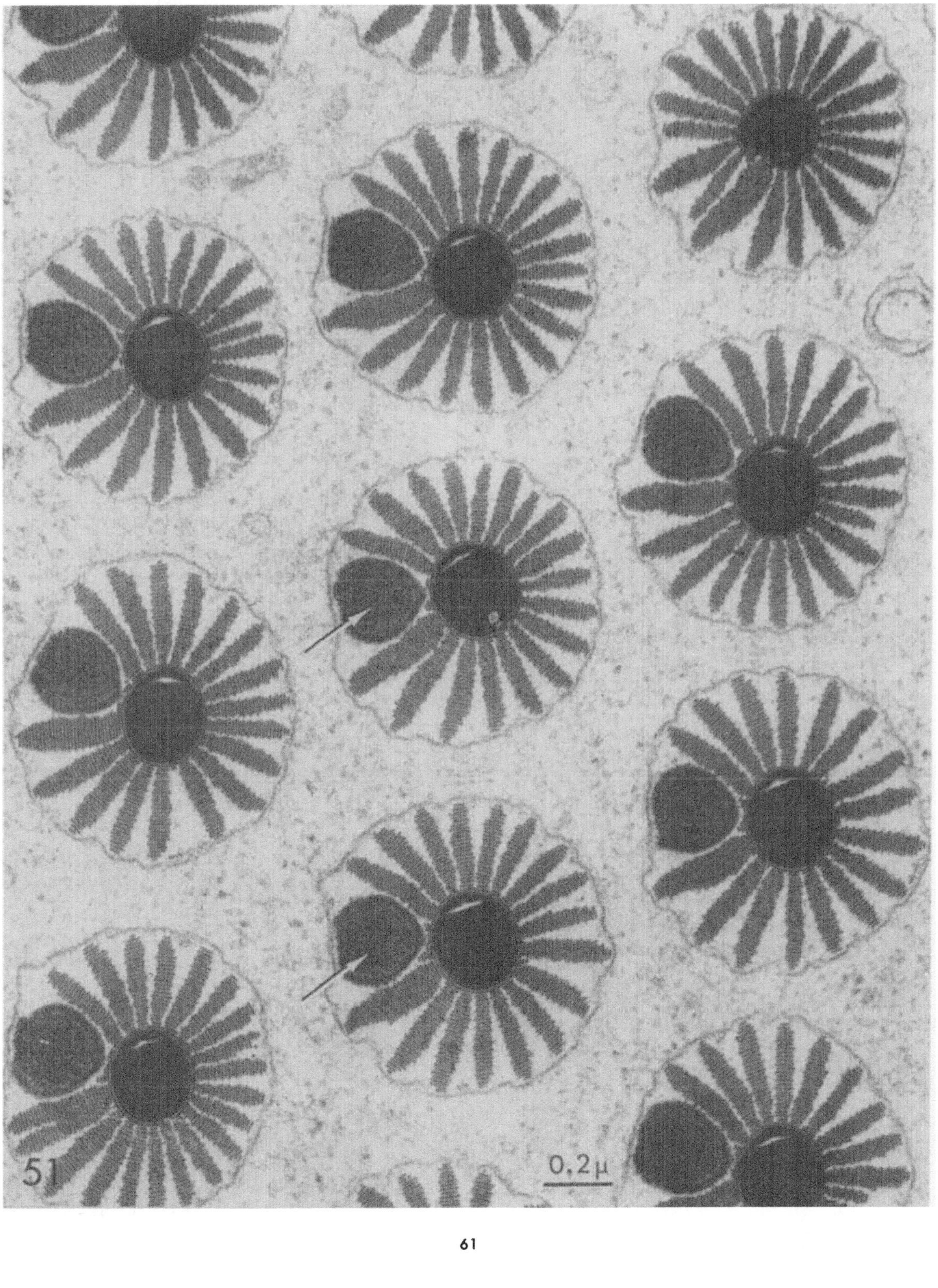
51
0.2μ

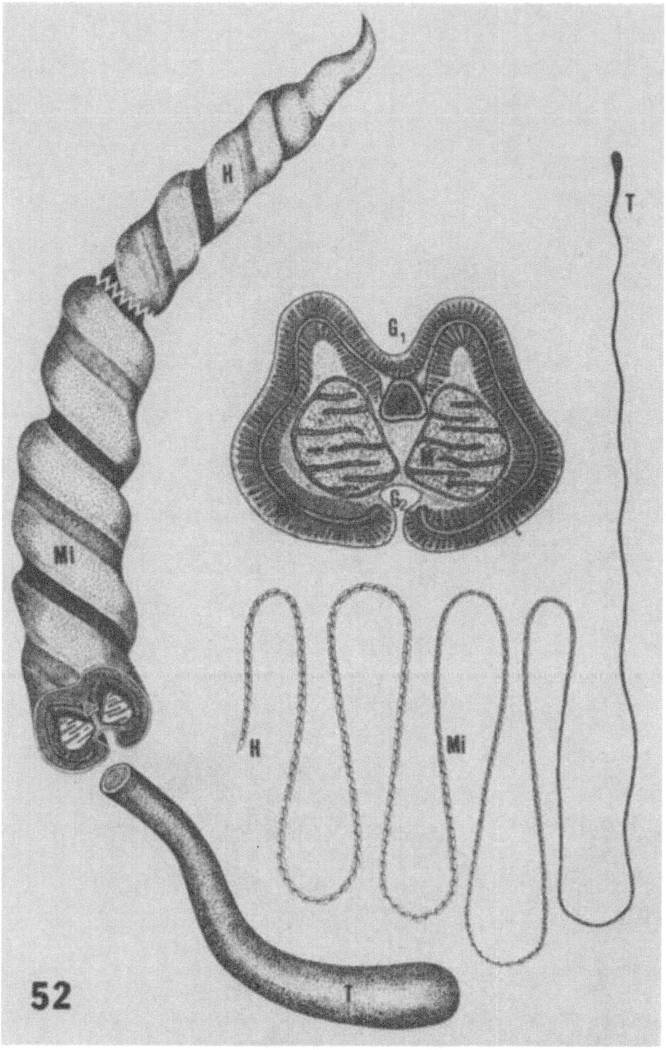


**Fig. 52.** The enlarged head (H), middle (Mi), and tail (T) segments of *Cypridopsis* sp. spermatozoon taken from complete spermatozoon shown in middle lower right of figure. Enlarged cross section taken from middle region of spermatozoon is shown in the middle portion of figure. Two grooves ($G_1$, $G_2$) present on surface of spermatozoon are staggered 180° with respect to each other and exhibit a pitch angle of approximately 45°. N, Nucleus; M, mitochondrion; MO, membranous organelle. Drawing and caption from B. Baccetti, "Comparative Spermatology." Academic Press, New York, 1970. Courtesy of Dr. J. F. Reger.

## EXCEPTIONAL SPERMATOZOA

Although spermatozoa from the large majority of animal species are long thin cells with an acrosome, nucleus, flagellum, and mitochondria, there are some very interesting exceptions among arthropods. Reger has described many of them (*118–125*). Isopod spermatozoa are long thin cells like other spermatozoa, but, instead of a flagellum, they contain a long thin rod composed at most levels of morphologically distinct central and peripheral material (Fig. 53). In longitudinal section, the peripheral portion appears striated and is morphologically similar to the columns of the connecting piece (*118, 119*). These spermatozoa are immotile. The spermatozoa of centipedes and millipedes are disc shaped, paired cells containing intranuclear mitochondria (*120, 121, 123*). The extensively studied spermatozoa of decapods have precise but strange shapes. Decapod spermatozoa are amoeboid cells containing an elaborate intranuclear acrosome (*5, 29, 79, 153*). The appearance of the chromatin in decapod spermatozoa, which lack mitochondria, is more typical of the appearance of procaryote than of eucaryote chromatin. Spermatozoa of nematodes are peculiar, aflagellate, amoeboid forms. Their chromatin never completely condenses, and they contain unique membrane-bound inclusions (*16, 54, 70*). One species has been described with a peculiar microtubular body (*80*). Spermatozoa of the ostracod *Cypridosis* have very elaborate mitochondria, which Reger (*121*) has suggested function as motile structures (Fig. 52). Gupta (*64*), however, has suggested that motility in ostracod spermatozoa is effected by a unique structure, the contractile band.

**Fig. 53.** Transverse section through spermatozoa of the marine isopod *Ligyda olfersii*.

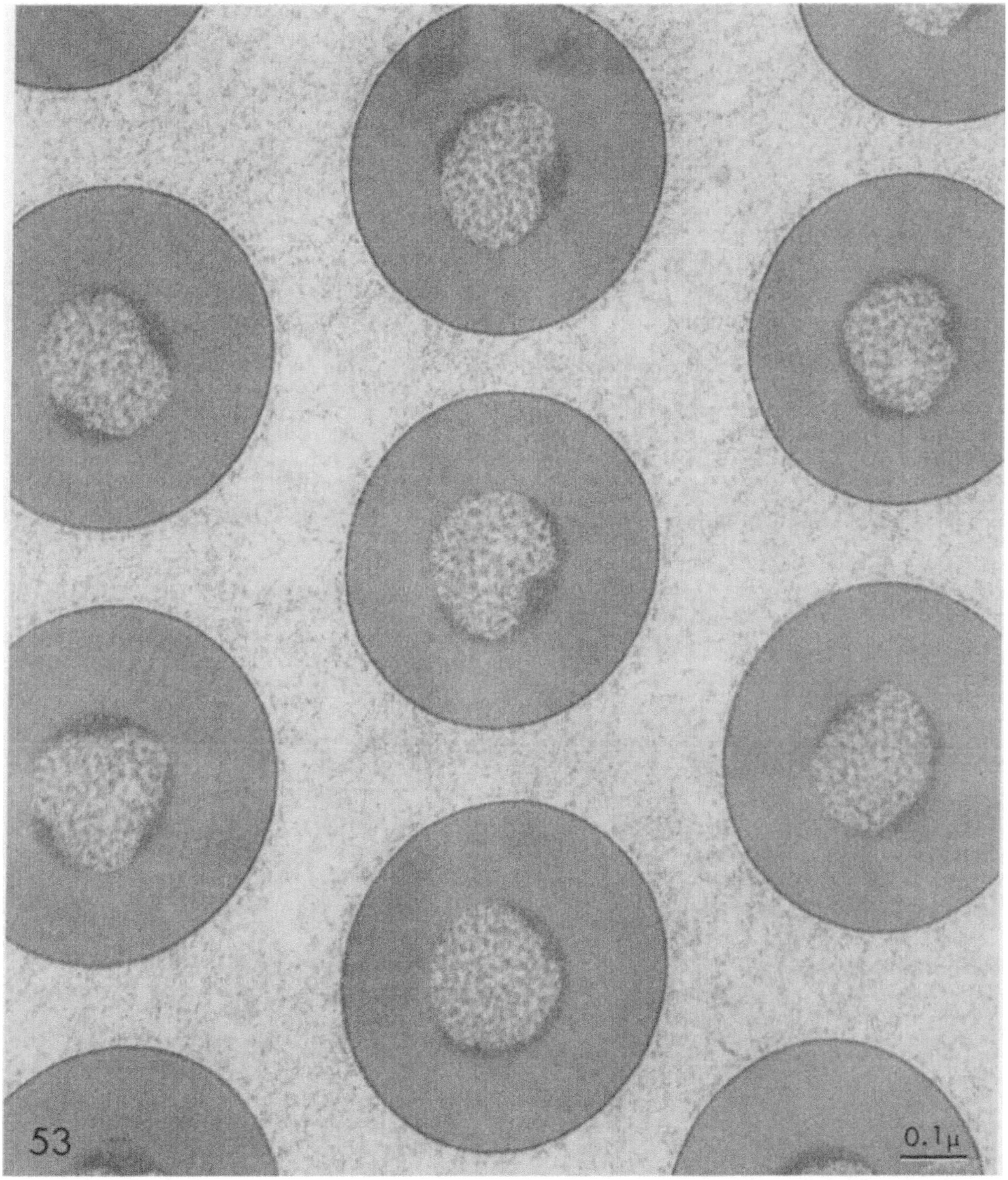
53
0.1μ

## REFERENCES


1. Alfert, M. (1956). Cytochemical differentiation of nuclear proteins during spermatogenesis in the salmon. *J. Biophys. Biochem. Cytol.* **2**, 109.
2. Anderson, W. A. (1968). Cytochemistry of sea urchin gametes. II. Ruthenium red staining of gametes of sea urchin. *J. Ultrastruct. Res.* **24**, 322.
3. Anderson, W. A. (1968). Cytochemistry of sea urchin gametes. I. Intramitochondrial localization of glycogen, glucose-6-phosphatase, and adenosine triphosphatase activity in spermatozoa of *Paracentrotus lividus*. *J. Ultrastruct. Res.* **24**, 398.
4. Anderson, W. A., and Curgy, J. (1969). RNA synthesis in nuclei and mitochondria during spermiogenesis *Lumbricus terrestris*. *J. Submicrosc. Cytol.* **1**, 25.
5. Anderson, W. A., and Ellis, R. A. (1966). Cytodifferentiation of the crayfish spermatozoon: Acrosome formation, transformation of mitochondria and development of microtubules. *Z. Zellforsch. Mikrosk. Anat.* **77**, 88.
6. Anderson, W. A., and Ellis, R. A. (1968). Acrosome morphogenesis in *Lumbricus terrestris*. *Z. Zellforsch. Mikrosk. Anat.* **85**, 398.
7. Anderson, W. A., and Personne, P. (1967). The fine structure of the neck region of spermatozoa of *Helix aspera*. *J. Microsc. (Paris)* **6**, 1033.
8. Anderson, W. A., and Personne, P. (1970). The localization of glycogen in the spermatozoa of various invertebrate and vertebrate species. *J. Cell Biol.* **44**, 29.
9. Anderson, W. A., Weisman, A., and Ellis, R. A. (1967). Cytodifferentiation during spermiogenesis in *Lumbricus terrestris*. *J. Cell Biol.* **32**, 11.
10. André, J. (1961). Sur quelques détails nouvellement connus de l'ultrastructure des organites vebratiles. *J. Ultrastruct. Res.* **5**, 86.
11. André, J. (1962). Contribution à la connaissance du chondriome. Etude de ses modifications ultrastructurales pendant la spermatogénèse. *J. Ultrastruct. Res. Res., Suppl.* **3**.
12. Baccetti, B., Dallai, R., and Rosati, F. (1969). The spermatozoon of Arthropoda. III. The lowest homometabolic insects. *J. Microsc. (Paris)* **8**, 233.
13. Baccetti, B., Bigliardi, E., and Rosati, F. (1971). The spermatozoon of Arthropoda. XIII. The cell surface. *J. Ultrastruct. Res.* **35**, 582.
14. Bainton, D. F., and Farquhar, M. G. (1966). Origin of granules in polymorphonuclear leukocytes. *J. Cell Biol.* **28**, 277.
15. Bairati, A. (1967). Struttura et ultrastruttura dell'apparato genitale maschile di *Drosophila melanogaster* Meig. I° JL Testicolo, Z. *Zellforsch. Mikrosk. Anat.* **76**, 56.
16. Beams, H. W., and Sekhon, S. S. (1972). Cytodifferentiation during spermiogenesis in *Rhabditis pellio*. *J. Ultrastruct. Res.* **38**, 511.
17. Bedford, J. M. (1966). Development of the fertilizing ability of spermatozoa in the epididymis of the rabbit. *J. Exp. Zool.* **163**, 319.
18. Bedford, J. M. (1968). Ultrastructural changes in the sperm head during fertilization in the rabbit. *Amer. J. Anat.* **123**, 329.
19. Bedford, J. M. (1970). Sperm capacitation and fertilization in mammals. *Biol. Reprod., Suppl.* **2**, 128.
20. Bedford, J. M., and Nicander, L. (1971). Ultrastructural changes in the acrosome and sperm membranes during maturation of spermatozoa in the testis and epididymis of rabbit and monkey. *J. Anat.* **108**, 527.
21. Behnke, O. (1967). Incomplete microtubules observed in blood platelets during microtubule polymerization. *J. Cell Biol.* **34**, 697.
22. Billard, R. (1970). Ultrastructure comparée de spermatozoides de quelques poissons Téléostéens. *In* "Comparative Spermatology" (B. Baccetti, ed.), pp. 71–79. Academic Press, New York.
23. Blandau, R. J., and Rummery, R. E. (1964). The relationship of swimming movements of epididymal spermatozoa to their fertilizing capacity. *Fert. Steril.* **15**, 571.
24. Bloch, D. P., and Brack, S. D. (1964). Evidence for the cytoplasmic synthesis of nuclear histone during spermiogenesis in the grasshopper *Chortophaga viridifasciata* (de Geer). *J. Cell Biol.* **22**, 327.
25. Bloch, D. P., and Hew, H. (1960). Schedule of spermatogenesis in the pulmonate snail *Helix aspera* with special reference to histone transition. *J. Biophys. Biochem. Cytol.* **7**, 515.
26. Boisson, C., and Mattei, X. (1965). Sur la spermiogénèse de Python sebae (Gmelin) étudie au microscope électronique. *C. R. Soc. Biol.* **159**, 1192.
27. Bonsdorff, C. H. von, and Talkka, A. (1965). The spermatozoon flagella in *Diphyllobothrium latum*. *Z. Zellforsch. Mikrosk. Anat.* **66**, 643.
28. Bradke, D. L. (1963). The origin and development of the manchette during spermatogenesis. *Anat. Rec.* **145**, 210.
29. Brown, G. G. (1966). Ultrastructural studies of sperm morphology and sperm. Egg interaction in the decapod *Callinectes sapidus*. *J. Ultrastruct. Res.* **14**, 425.
30. Burgos, M. H., and Fawcett, D. W. (1955). Studies on the fine structure of the mammalian testis. I. Differentiation of the spermatids in the cat (*Felis domestica*). *J. Biophys. Biochem. Cytol.* **1**, 287.
31. Cameron, M. L. (1965). Some details of the ultrastructure in the development of flagellar fibers of the tenebrio sperm. *Can. J. Zool.* **43**, 1005.
32. Cantacuzene, A. M. (1970). L'annexe centriolaire du spermatozoide des insectes. *In* "Comparative Spermatology" (B. Baccetti, ed.), pp. 553–563. Academic Press, New York.
33. Christensen, A. K. (1961). Fine structure of an unusual spermatozoon in the flatworm *Plagiostomum*. *Biol. Bull.* **121**, 416.
34. Clark, A. W. (1967). Some aspects of spermiogenesis in a lizard. *Amer. J. Anat.* **121**, 369.
35. Clermont, Y. (1960). Cycle of the seminiferous epithelium of the guinea pig. *Fert. Steril.* **6**, 563.
36. Cohen, W. D., and Gottlieb, T. (1971). C-microtubules in isolated mitotic spindles. *J. Cell Sci.* **9**, 603.
36a. Colwin, L. H., and Colwin, A. L. (1967). Membrane fusion in relation to sperm–egg association. *In* "Fertilization" (C. B. Metz and A. Monroy, eds.), Vol. 1, p. 295. Academic Press, New York.
37. Comings, D. E., and Okada, T. A. (1972). The chromatoid body in mouse spermatogenesis: Evidence that it may be formed by the extension of nucleolar components. *J. Ultrastruct. Res.* **39**, 15.
38. Das, C. C., Kaufman, B. P., and Gay, H. (1964). Histone–protein transitions in *Drosophila melanogaster*. I. Changes during spermatogenesis. *Exp. Cell Res.* **35**, 507.
39. Dietert, S. E. (1966). Fine structure of the formation and fate of the residual bodies of mouse spermatozoa

with evidence for the participation of lysosomes. *J. Morphol.* **120,** 317.
40. Dym, M., and Fawcett, D. W. (1971). Further observations on the number of spermatogonia, spermatocytes, and spermatids connected by intercellular bridges in the mammalian testis. *Biol. Reprod.* **4,** 195.
41. Eddy, E. M. (1970). Cytochemical observations on the chromatoid body of male germ cells. *Biol. Reprod.* **2,** 114.
42. Fawcett, D. W. (1968). The topographical relationship between the plane of the central pair of flagellar fibrils and the transverse axis of the head in guinea pig spermatozoa. *J. Cell Sci.* 3, 187.
43. Fawcett, D. W. (1971). A comparative view of sperm ultrastructure. *Biol. Reprod.* **2,** 90.
44. Fawcett, D. W. (1972). Observations on cell differentiation and organelle continuity in spermatogenesis. *In* "International Symposium on the Genetics of the Spermatozoon" (R. A. Beatty and S. Gluecksonn-Waelsh, eds.), p. 37.
45. Fawcett, D. W., Anderson, W. A., and Phillips, D. M. (1971). Morphogenetic factors inyuencing the shape of the sperm head. *Develop. Biol.* **26,** 220.
46. Fawcett, D. W., Eddy, E. M., and Phillips, D. M. (1970). Observations on the fine structure and relationships of the chromatoid body in mammalian spermatogenesis. *Biol. Reprod.* **2,** 129.
47. Fawcett, D. W., and Hollenberg, R. (1963). Changes in the acrosome of guinea pig spermatozoa during passage through the epididymis. *Z. Zellforsch. Mikrosk. Anat.* **60,** 276.
48. Fawcett, D. W., and Ito, S. (1965). The fine structure of bat spermatozoa. *Amer. J. Anat.* **116,** 567.
49. Fawcett, D. W., and Phillips, D. M. (1969). The fine structure and development of the neck region of the mammalian spermatozoon. *Anat. Rec.* **165,** 153.
50. Fawcett, D. W., and Phillips, D. M. (1969). Observations on the release of spermatozoa and on changes in the head during passage through the epididymis. *J. Reprod. Fert., Suppl.* **6,** 405.
51. Fawcett, D. W., and Phillips, D. M. (1970). Recent observations on the ultrastructure and development of the mammalian spermatozoon. *In* "Comparative Spermatology" (B. Baccetti, ed.), pp. 13–28. Academic Press, New York.
52. Felix, K., Fischer, H., and Krekels, A. (1956). Protamines and nucleoprotamines. *Progr. Biophys. Biophys. Chem.* **6,** 6.
53. Ferraguti, M., and Lanzavecchia, G. (1971). Morphogenetic effects of microtubules. I. Spermiogenesis in *Annelida tubificidae. J. Submicrosc. Cytol.* **3,** 121.
54. Foor, W. E. (1968). Zygote formation in *Ascaris lumbricoides* (Nematoda). *J. Cell Biol.* **39,** 119.
55. Furier, P. (1970). Sperm morphology in some reptiles: Squamata and Chelonia. *In* "Comparative Spermatology" (B. Baccetti, ed.), pp. 115–131. Academic Press, New York.
56. Gaddum, P. (1968). Sperm maturation in the male reproductive tract: Development of motility. *Anat. Rec.* **161,** 471.
57. Gibbons, I. R., and Grimstone, A. V. (1960). On flagellar structure in certain flagellates. *J. Biophys. Biochem. Cytol.* **7,** 697.
58. Giusti, F. (1969). The spermatozoon of a fresh-water prosobranch mollusc. *J. Submicrosc. Cytol.* **1,** 263.
59. Gordon, M. (1972). The distal centriole in guinea pig spermiogenesis. *J. Ultrastruct. Res.* **39,** 364.
60. Gordon, M., and Barrnett, R. J. (1967). Fine structural cytochemical localizations of phosphatase activities of rat and guinea pig. *Exp. Cell Res.* **48,** 395.
61. Gordon, M., and Bensch, K. G. (1968). Cytochemical differentiation of the guinea pig sperm flagellum with phosphoturgastic acid. *J. Ultrastruct. Res.* **24,** 33.
62. Gray, J. (1958). The movement of spermatozoa of the bull. *J. Exp. Biol.* **35,** 96.
63. Grimstone, A. V., and Cleveland, L. R. (1965). The fine structure and function of the contractile axostyles of certain flagellates. *J. Cell Biol.* **24,** 387.
64. Gupta, B. L. (1968). Aspects of motility in the non-flagellate spermatozoa of freshwater ostracods. *Symp. Soc. Exp. Biol.* **22,** 138.
65. Hamilton, D. W., and Fawcett, D. W. (1968). Unusual features of the neck and middle-piece of snake spermatozoa. *J. Ultrastruct. Res.* **23,** 81.
66. Henderson, S. A. (1964). RNA synthesis during male meiosis and spermiogenesis. *Chromosoma* **15,** 345.
67. Hepler, P. K., McIntosh, R. J., and Cleland, S. (1970). Intermicrotubule bridges in the mitotic spindle apparatus. *J. Cell Biol.* **45,** 438.
68. Hughes-Schrader, S. (1948). Cytology of coccids (Coccoidea–Homoptera). *Advan. Genet.* **2,** 127.
69. Inoue, S., and Hidemi, S. (1962). Arrangement of DNA in living sperm. A. biophysical analysis. *Science* **136,** 1122.
70. Jamaur, M. P. (1966). Studies of spermiogenesis in a nematode *Nippostronglylus brasiliensis. J. Cell Biol.* **31,** 381.
71. Kaye, J. S. (1958). Changes in the fine structure of mitochondria during spermatogenesis. *J. Morphol.* **102,** 347.
72. Kaye, J. S. (1962). Acrosome formation in the house cricket. *J. Cell Biol.* **12,** 411.
73. Kaye, J. S., and Kaye, R. M. (1966). The fine structure and chemical composition of nuclei during spermiogenesis in the house cricket. I. Initial stages of differentiation and the loss of nonhistone protein. *J. Cell Biol.* **31,** 159.
74. Kessel, R. G. (1966). The association between microtubules and nuclei during spermiogenesis in the dragonfly. *J. Ultrastruct. Res.* **16,** 293.
75. Kessel, R. G. (1967). An electron microscope study of spermiogenesis in the grasshopper with particular reference to the development of microtubular systems during differentiation. *J. Ultrastruct. Res.* **18,** 677.
76. Kessel, R. G. (1970). Spermiogenesis in the dragonfly with special reference to a consideration of the mechanisms involved in the development of cellular asymmetry. *In* "Comparative Spermatology" (B. Baccetti, ed.), pp. 531–552. Academic Press, New York.
77. Kiefer, B. I. (1966). Ultrastructural abnormalities in developing sperm of X/O *Drosophila melanogaster. Genetics* **54,** 1441.
78. Koehler, J. K. (1970). A freeze-etching study of rabbit spermatozoa with particular reference to head structures. *J. Ultrastruct. Res.* **33,** 598.
79. Langreth, S. G. (1969). Spermiogenesis in *Cancer* crabs. *J. Cell Biol.* **43,** 575.
80. Lee, D. L., and Anya, A. O. (1967). The structure and development of the spermatozoon of *Aspiculuris tetraptera* (Nematoda). *J. Cell Sci.* **2,** 537.
81. Leuchtenberger, C., and Schrader, F. (1950). The chemical nature of the acrosome in the male germ cells. *Proc. Nat. Acad. Sci. U.S.* **36,** 677.
82. Lommen, M. A. J. (1970). Development of the ring

during spermiogenesis of a salamander. *J. Microsc. (Paris)* **9,** 785.
83. Longo, F. J., and Anderson, E. (1970). Structural and cytochemical features of the sperm of the Cephalopod *Octopus bimaculatus*. *J. Ultrastruct. Res.* **32,** 94.
84. MacGregor, H. C., and Kezer, J. (1971). The chromosomal localization of a heavy satellite DNA in the testis of *Plethodon c. cinereus*. *Chromosoma* **33,** 167.
84a. MacGregor, H. C., and Walker, M. H. (1973). The arrangement of chromosomes in nuclei of sperm of plethodontid salamanders. *Chromosoma* **40,** 243.
85. Maillet, P. L., and Gouranton, A. (1965). Sur l'expulsion de l'acide ribonucléique nucléaire par les spermatides de *Philaenus spumarius*. *L. C. R. Acad. Sci.* **261,** 1417.
86. Mancini, R. E., Alonso, A., Barquet, J., Alvarez, B., and Nemirovsky, M. (1964). Histoimmunological localization of hyaluronidase in the bull testis. *J. Reprod. Fert.* **8,** 325.
87. Mattei, C., Mattei, X., and Manfreidi, J. (1972). Electron microscope study of the spermiogenesis of *Streptopelia roseogrisea*. *J. Submicrosc. Cytol.* **4, 57.**
88. Mattei, X. (1970). Spermiogénèse comparée des poissons. *In* "Comparative Spermatology" (B. Baccetti, ed.), pp. 57–70. Academic Press, New York.
89. McIntosh, J. R., and Porter, K. R. (1967). Microtubules in the spermatids of the domestic fowl. *J. Cell Biol.* **35,** 153.
90. Melzel, S. (1970). Cytochemical studies on nonspecific esterases of bull spermatozoa. *Biol. Reprod.* **2,** 18.
91. Meyer, G. F. (1964). Die parakristallinen Körper in den Spermienschwanzen von *Drosophila*. *Z. Zellforsch. Mikrosk. Anat.* **62,** 762.
92. Monesi, V. (1965). Synthetic activities during spermatogenesis in the mouse. RNA and protein synthesis. *Exp. Cell Res.* **39,** 197.
93. Moses, M. J. (1970). Spermiogenesis in an Iceryine coccid, *Steatococcus tuberaculatus* Morrison. An electron microscopic study. *Chromosoma* **30,** 373.
94. Muckenthaler, F. A. (1964). Autoradiographic study of nucleic acid synthesis during spermatogenesis in the grasshopper, *Melanoplus differentialis*. *Exp. Cell Res.* **35,** 531.
95. Nelson, L. (1958). Cytochemical studies with the electron microscope. I. Adenosinetriphosphate in rat spermatozoa. *Biochim. Biophys. Acta* **27,** 634.
96. Nicander, L. (1970). Comparative studies on the fine structure of vertebrate spermatozoa. *In* "Comparative Spermatology" (B. Baccetti, ed.), pp. 47–56. Academic Press, New York.
97. Olds, P. J. (1971). Effect of the T locus on sperm ultrastructure in the house mouse. *J. Anat.* **107,** 31.
98. Pedersen, H. (1970). Ultrastructure of the ejaculated human sperm. *In* "Comparative Spermatology" (B. Baccetti, ed.), pp. 133–142. Academic Press, New York.
99. Phillips, D. M. (1966). Observations on spermiogenesis in the fungus gnat *Sciara coprophila*. *J. Cell Biol.* **30,** 477.
100. Phillips, D. M. (1966). Fine structure of *Sciara coprophila* sperm. *J. Cell Biol.* **30,** 499.
101. Phillips, D. M. (1969). Exceptions to the prevailing pattern of tubules $(9 + 9 + 2)$ in the sperm flagella of certain insect species. *J. Cell Biol.* **40,** 28.
102. Phillips, D. M. (1970). Development of spermatozoa in the woolly opossum with special reference to the shaping of the sperm head. *J. Ultrastruct. Res.* **33,** 369.
103. Phillips, D. M. (1970). Insect flagellar tubule patterns: Theme and variations. *In* "Comparative Spermatology" (B. Baccetti, ed.), pp. 263–274. Academic Press, New York.
104. Phillips, D. M. (1970). Insect sperm: Their structure and morphogenesis. *J. Cell Biol.* **44,** 243.
105. Phillips, D. M. (1970). Ultrastructure of spermatozoa of the woolly opossum *Caluromys philander*. *J. Ultrastruct. Res.* **33,** 381.
106. Phillips, D. M. (1971). Morphogenesis of the lacinate appendages of *Lepidopteran* spermatozoa. *J. Ultrastruct. Res.* **34,** 567.
107. Phillips, D. M. (1972). Comparative analysis of mammalian sperm motility. *J. Cell Biol.* **53,** 561.
108. Phillips, D. M. (1972). Substructure of the mammalian acrosome. *J. Ultrastruct. Res.* **38,** 591.
109. Phillips, D. M., and Swift, H. (1965). Cytoplasmic fine structure of *Sciara* salivary glands. I. Secretion. *J. Cell Biol.* **27,** 395.
109a. Piko, L. (1969). Gamete structure and sperm entry in mammals. *In* "Fertilization" (C. B. Metz and A. Monroy, eds.), p. 325, Academic Press, New York.
110. Plattner, H. (1971). Bull spermatozoa: A re-investigation by freeze-etching using widely different cryofixation procedures. *J. Submicrosc. Cytol.* **3,** 19.
111. Ploën, L. (1971). A scheme of rabbit spermateleosis based upon electron microscopial observations. *Z. Zellforsch. Mikrosk. Anat.* **115,** 553.
112. Pollister, A. W., and Mirsky, A. E. (1946). The nucleoprotamine of trout sperm. *J. Gen. Physiol.* **30,** 101.
113. Pratt, S. A. (1968). An electron microscope study of nebenkern formation and differentiation in spermatids of *Murgantia histrionica* (Hemiptera, Pentatomidae). *J. Morphol.* **126,** 31.
114. Rattner, J. B. (1972). Nuclear shaping in marsupial spermatids. *J. Ultrastruct. Res.* **32,** 316.
115. Rattner, J. B., and Brinkley, B. R. (1970). Ultrastructure of mammalian spermiogenesis. I. A tubular complex in developing sperm of the cottontop marmoset *Sequinus oedipus*. *J. Ultrastruct. Res.* **32,** 316.
116. Rattner, J. B., and Brinkley, B. R. (1971). Ultrastructure of mammalian spermiogenesis. II. Elimination of the nuclear membrane. *J. Ultrastruct. Res.* **36,** 1.
117. Rattner, J. B., and Brinkley, B. R. (1972). Ultrastructure of mammalian spermiogenesis. III. The organization and morphogenesis of the manchette during rodent spermiogenesis. *J. Ultrastruct. Res.* **41,** 209.
118. Reger, J. F. (1964). A study on the fine structure of developing spermatozoa from the isopod *Asellus militaris* (Hay). *J. Micros. (Paris)* **3,** 559.
119. Reger, J. F. (1966). A comparative study of the fine structure of developing spermatozoa in the isopod, *Oniscus asellus,* and the amphipod, *Orchestoidea* sp. *Z. Zellforsch. Mikrosk. Anat.* **75,** 579.
120. Reger, J. F. (1969). A fine structure study on spermiogenesis in the arachnid *Leiobunum* sp. (Phalangida, Harvestman). *J. Ultrastruct. Res.* **28,** 422.
121. Reger, J. F. (1970). Some aspects of the fine structure of filiform spermatozoa (Ostracod, *Cypridopsis* sp.) lacking tubule sub-structure. *In* "Comparative Spermatology" (B. Baccetti, ed.), pp. 237–246. Academic Press, New York.
122. Reger, J. F. (1971). Studies on the fine structure of spermatids and spermatozoa from the millipede *Spirobolus* sp. *J. Submicrosc. Cytol.* **3,** 33.

123. Reger, J. F., and Cooper, D. P. (1968). Studies on the fine structure of spermatids and spermatozoa from the millipede *Polydesmus* sp. *J. Ultrastruct. Res.* **23**, 60.
124. Reger, J. F., and Florendo, N. T. (1969). Studies on motile, non-tubule containing filiform spermatozoa of the ostracod *Cypridopsis* sp. I. Spermiogenesis. *J. Ultrastruct. Res.* **28**, 235.
125. Reger, J. F., and Florendo, N. T. (1969). Studies on motile, non-tubule containing filiform spermatozoa of the ostracod *Cypidopsis* sp. II. Mature spermatozoa. *J. Ultrastruct. Res.* **28**, 250.
126. Riemann, J. G. (1970). Metamorphosis of sperm of the cabbage looper, *Trichoplusia ni*, during passage from the testes to the female spermatheca. *In* "Comparative Spermatology" (B. Baccetti, ed.), pp. 321–332. Academic Press, New York.
127. Rikmenspoel, R. (1965). Tail movement of bull spermatozoa. Observations and model calculations. *Biophys. J.* **5**, 365.
128. Robison, W. G. (1966). Microtubules in relation to the motility of a sperm syncytium in an armored scale insect. *J. Cell Biol.* **29**, 251.
129. Robison, W. G. (1970). Unusual arrangements of microtubules in relation to mechanisms of sperm movement. *In* "Comparative Spermatology" (B. Baccetti, ed.), pp. 311–320. Academic Press, New York.
130. Romrell, L. J., Stanley, H. P., and Bowman, J. T. (1972). Genetic control of spermiogenesis in *Drosophila melanogaster:* An autosomal mutant (ms(2)3R) demonstrating failure of meiotic cytokinesis. *J. Ultrastruct. Res.* **38**, 563.
131. Ross, J., and Robison, W. G. (1969). Unusual microtubular patterns and three-dimensional movement of mealybug sperm and sperm bundles. *J. Cell Biol.* **40**, 426.
132. Roth, L. E., and Shagenaka, Y. (1970). Microtubules in the Heliozoan axopodium. II. Rapid degradation by cupric and nickelous ions. *J. Ultrastruct. Res.* **31**, 356.
133. Sandoz, D. (1970). Etude ultrastructurale et cytochimique de la formation de l'acrosome du *Discogluosse* (amphibien anoure). *In* "Comparative Spermatology" (B. Baccetti, ed.), pp. 93–114. Academic Press, New York.
134. Satir, P. (1965). Studies on cilia. II. Examination of the distal region or the ciliary shaft and the role of the filaments in motility. *J. Cell Biol.* **26**, 805.
135. Shapiro, J. E., Hershenov, B. R., and Tulloch, G. S. (1961). The fine structure of Haematolaechus spermatozoon tails. *J. Biophys. Biochem. Cytol.* **9**, 211.
136. Shay, J. W. (1972). Electron microscope studies of spermatozoa of *Rhynchosciara* sp. I. Disruption of microtubules by various treatments. *J. Cell Biol.* **54**, 598.
137. Shoup, J. R. (1967). Spermiogenesis in wild type and in a male sterility mutant of *Drosophila melanogaster. J. Cell Biol.* **32**, 663.
138. Silveira, M., and Porter, K. R. (1964). The spermatozoids of flatworms and their microtubular systems. *Protoplasma* **59**, 240.
139. Stanley, H. P. (1969). An electron microscope study of spermiogenesis in the teleost fish *Oligocottus maculosus. J. Ultrastruct. Res.* **27**, 230.
140. Stanley, H. P. (1971). Fine structure of spermiogenesis in the elasmobranch fish *Squalus suckley.* II. Late stages of differentiation and structure of mature spermatozoon. *J. Ultrastruct. Res.* **36**, 103.
141. Taylor, J. H. (1964). The arrangement of chromosomes in the mature sperm of grasshopper. *J. Cell Biol.* **21**, 286.
142. Tucker, J. B. (1968). Fine structure and function of the cytopharyngeal basket in the ciliate *Nassula. J. Cell Sci.* **3**, 493.
143. Tucker, J. B. (1972). Microtubule-arms and propulsion of food particles inside a large feeding organelle in the ciliate *Phasolodon vorticella. J. Cell Sci.* **10**, 883.
144. Utakoji, T. (1966). Crondogy of nucleic acid synthesis in meiosis of the male Chinese hamster. *Exp. Cell Res.* **42**, 585.
145. Virkki, N. (1969). Sperm bundles and phylogenesis. *Z. Zellforsch. Mikrosk. Anat.* **101**, 13.
146. Warner, F. (1971). Spermatid differentiation in the blowfly *Sarcophaga bullata* with particular reference to flagellar morphogenesis. *J. Ultrastruct. Res.* **35**, 210.
147. Werner, G. (1966). Untersuchungen über die Spermiogenese bei einem Laufkafer *Carabus catenulatus* scop., und der Skorpion-wassersanze, *Nepa rubra* L. *Z. Zellforsch. Mikrosk. Anat.* **73**, 576.
148. Werner, G. (1970). On the development and structure of the neck in urodele sperm. *In* "Comparative Spermatology" (B. Baccetti, ed.), pp. 85–92. Academic Press, New York.
149. Wilson, E. B. (1928). "The Cell in Development and Heredity." Macmillan, New York.
150. Woolley, D. M. (1970). The midpiece of the mouse spermatozoon: Its form and development as seen by surface replication. *J. Cell Sci.* **6**, 865.
151. Woolley, D. M., and Beatty, R. A. (1967). Inheritance of midpiece length in mouse spermatozoa. *Nature (London)* **215**, 94.
151a. Yanagimachi, R., Noda, Y. D., Fujimoto, M., and Nicolson, G. L. (1972). The distribution of negative surface charges on mammalian spermatozoa. *Amer. J. Anat.* **135**, 497.
152. Yanagimachi, R., and Teichman, R. J. (1972). Cytochemical demonstration of acrosomal proteinase in mammalian avian spermatozoa by a silver proteinate method. *Biol. Reprod.* **6**, 87.
153. Yasuzumi, G., and Lee, K. (1966). Spermatogenesis in animals as revealed by electron microscopy. XVI. The microtubular structure and sites of thiamine pyrophosphatase activity in premature sperm of the Japanese crayfish. *Z. Zellforsch. Mikrosk. Anat.* **73**, 384.
154. Yasuzumi, G., Matano, Y., Asai, T., Nagasaks, M., and Yasuzumi, F. (1971). Spermatogenesis in animals as revealed by electron microscopy. XII. Development of nuclei and cytoplasmic microtubules in the grasshopper spermatids. *Z. Zellforsch. Mikrosk. Anat.* **115**, 243.
155. Yasuzumi, G., and Sugioka, T. (1971). Spermatogenesis in animals as revealed by electron microscopy. XXI. Microkaryosomes and microtubules appearing during spermiogenesis of the lovebird *Uroloncha striata* var. *domestica* Flower. *Z. Zellforsch. Mikrosk. Anat.* **114**, 451.
156. Young, W. C. (1931). A study of the function of the epididymis. III. Functional changes undergone by spermatozoa during their passage through the epididymis and vas deferens in the guinea pig. *Brit. J. Exp. Biol.* **8**, 151.
157. Zamboni, L., and Stefanini, M. (1971). The fine structure of the neck of mammalian spermatozoa. *Anat. Rec.* **169**, 121.

## INDEX

Printed in France by Amazon
Brétigny-sur-Orge, FR

51786481R00045